T0282329

CAMBRIDGE STUDIES
IN MATHEMATICAL BIOLOGY: 8
Editors
C. CANNINGS
Department of Probability and Statistics, University of Sheffield,
Sheffield, U.K.
F. C. HOPPENSTEADT
College of Natural Sciences, Michigan State University,
East Lansing, Michigan, U.S.A.
L. A. SEGEL
Weizmann Institute of Science, Rehovot, Israel

INTRODUCTION TO
THEORETICAL NEUROBIOLOGY
Volume 2

HENRY C. TUCKWELL

Monash University

Introduction to theoretical neurobiology: volume 2 nonlinear and stochastic theories

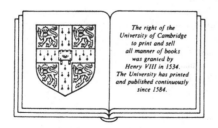

The right of the
University of Cambridge
to print and sell
all manner of books
was granted by
Henry VIII in 1534.
The University has printed
and published continuously
since 1584.

CAMBRIDGE UNIVERSITY PRESS

Cambridge

New York   New Rochelle

Melbourne   Sydney

CAMBRIDGE UNIVERSITY PRESS
Cambridge, New York, Melbourne, Madrid, Cape Town, Singapore, São Paulo

Cambridge University Press
The Edinburgh Building, Cambridge CB2 2RU, UK

Published in the United States of America by Cambridge University Press, New York

www.cambridge.org
Information on this title: www.cambridge.org/9780521352178

First published 1988
This digitally printed first paperback version 2005

A catalogue record for this publication is available from the British Library

Library of Congress Cataloguing in Publication data
Tuckwell, Henry C. (Henry Clavering), 1943–
Introduction to theoretical neurobiology.
(Cambridge studies in mathematical biology; 8)
Includes bibliographies and index.
Contents: v. 1. Linear cable theory and dendritic
structure—v. 2. Nonlinear and stochastic theories.
1. Neurons—Mathematical models. 2. Neural
transmission—Mathematical models. I. Title.
II. Series.
QP361.T76   1988   559'.0188   87-25690

ISBN-13 978-0-521-35217-8 hardback
ISBN-10 0-521-35217-7 hardback

ISBN-13 978-0-521-01932-3 paperback
ISBN-10 0-521-01932-X paperback

To My Sister, Laurica

CONTENTS

PREFACE

In Volume 1, *Linear Cable Theory and Dendritic Structure*, the classical theory of membrane potentials, the Lapicque model neuron, and linear cable theory were introduced. The subthreshold responses described were always linear and the mathematical theory relatively simple.

In this volume, *Nonlinear and Stochastic Theories*, more complicated models are considered with concomitant increases in mathematical complexity. The principal new elements are nonlinear systems of reaction–diffusion systems and probability and stochastic processes. However, the required mathematics is developed in the text. The material also involves numerical methods of solution of partial differential equations and the statistical analysis of point processes. It is expected that this volume would be suitable for a second one-term course in quantitative neurophysiology with Volume 1 as a prerequisite.

Chapter 7 introduces reversal potentials for ionic currents into the cable equation with emphasis on those that arise during synaptic transmission. A derivation of the cable equation with synaptic reversal potentials is given as there did not appear to be one in the literature. The space-clamped problems are dealt with first, for the sake of completeness of presentation, and are used to explore the effects of relative timing of excitation and inhibition. Using the Green's function technique, exact solutions are then obtained for the full cable equation with periodic synaptic excitation. The interaction between excitatory and inhibitory inputs is investigated and leads to the definition of amplification factors, which can be found exactly in some cases. Shunting inhibition, relevant to theories of direction-sensitive retinal cells, is discussed. Further theoretical investigations are made of the arrival of synaptic input when a steady current is injected at the soma, EPSPs in response to sustained conductance charges. In the latter, exact results are given for the infinite cable and a solution is obtained by successive substitution in a Volterra equation in the case of finite cables.

Chapter 8 contains the theory of action-potential generation based on the Hodgkin–Huxley equations. The first three sections introduce the physical concepts and formulate the equations. The space-clamped equation and the traveling-wave equation are developed. There follows a general discussion of reaction–diffusion systems of equations and a method for solving them numerically is presented in detail. Attention is then directed to various forms of the Fitzhugh–Nagumo equations whose solutions mimic those of the Hodgkin–Huxley system. A summary of the relevant mathematical theory is given, including that of repetitive firing and bifurcation theory. A sketch is made of recent progress with the analysis of the Hodgkin–Huxley equations. The rest of the chapter concerns action-potential propagation in structures other than a purely cylindrical axon. Topics include the myelinated nerve and the Frankenhaeuser–Huxley equations, the effects of various geometries including branching, the accumulation of potassium in the extracellular space, and interactions between fibers in a nerve bundle.

Chapter 9 is devoted entirely to stochastic models of neural activity. Some of the stochastic phenomena encountered in observations on nerve cells are explained and then a brief survey of the concepts of elementary probability is given, mainly to explain the notation. After a brief look at the quantum hypothesis and its successful application via compound Poisson random variables, the theory of the relevant random processes is explained. These processes include Poisson processes and Weiner processes, both of which have been employed as approximate nerve-cell models. The central problem, that of the determination of the distribution of the interspike interval, is a first-passage (or exit) time problem. The exact known (classical) results are derived for several processes. A brief excursion is made into the general theory of Markov processes with emphasis on those describable by Itô-type stochastic differential equations. These include both diffusion processes and discontinuous processes. This leads naturally to Stein's model and its generalizations and this theory is treated at length. The natural diffusion approximation for Stein's model is the Ornstein–Uhlenbeck process and this is considered next. The well-known first-passage time results are derived for this process and some results that reveal the accuracy of the approximation are given. Mention is made of the rather technical mathematical developments in this area, in particular, the recent weak-convergence results. The last section of Chapter 9 is devoted to stochastic cable theory, which involves the currently interesting topic of stochastic partial differential equations. Thus theoretical neurobiology comes in contact with modern applied mathematics in deterministic (Chapter 8) and stochastic frameworks.

Chapter 10, the last in the present volume, addresses the statistical analysis of neuronal data with emphasis on spike trains. The standard renewal model is first presented with its relevant properties. There follows a bevy of statistical procedures, which include estimating the parameters of standard distributions, testing and comparing distributions, testing for stationarity, independence, for Poisson and renewal processes. Some generalizations and departures from the standard renewal model are analyzed and then the interesting problem of parameter estimation for diffusion processes is addressed. An attempt is made to classify all known ISI distributions into 10 basic kinds. This is followed by an examination of the theory of poststimulus time histograms and lastly by the techniques that here have been employed to analyze the ongoing activity of several neurons.

Again various parts have been written in various places. Chapters 7, 8, and 9 were mostly completed during 1984–5 at Monash University. Chapter 10 was started at the University of Washington in 1986 and finished at UCLA in 1987. Dr. James Koziol made many helpful comments on Chapter 10, and Dr. John B. Walsh and Dr. Geoffrey Watterson offered useful remarks concerning the earlier material. I am grateful to Monash University for the leaves of absence that have enabled me to finish this work and to my various collaborators with whom some of this work, especially in Chapter 9, has been a collaborative effort. In particular, Drs. George Bluman, Davis Cope, Floyd Hanson, and Frederic Wan. For technical assistance I thank Ian Coulthard, Babette Dalton, Jean Sheldon, Nereda Shute, Avis Williams and Barbara Young.

Los Angeles, March 1987 Henry C. Tuckwell

7

Subthreshold response to synaptic input

7.1   Introduction: cable equation with ionic currents

The previous three chapters were devoted to solving the
equation $V_t = V_{xx} - V + I$ in order to find the subthreshold responses
of nerve cylinders and neurons with dendritic trees to various input
currents. Many results were obtained by using the Green's function
method of solution. Among those results was the response to a local
input current of the form

$$f(t) = bte^{-at}. \tag{7.1}$$

When a is large this gives a current pulse of short duration, which has
been employed as an *approximation* to the current that occurs at the
site of a synapse when that synapse is activated. Although this
approach is useful for studying the response to a single synaptic event,
or events, at the same time at different synapses when the cell is
initially at resting state, it does not give an accurate description of the
dependence of the response on the prior state of excitation and it does
not provide, in general, an accurate description of the interaction
between synaptic stimuli.

Recall that when a synapse, whose activation is mediated by release
of transmitter from the presynaptic terminal, is active, the permeabilities of the postsynaptic membrane to various ions undergo local
changes. However, the mathematical treatment of ionic fluxes through
the various specialized ion channels under the combined effects of
electrical and diffusive forces is too difficult. In order to overcome
this problem, a heuristic approach is adopted. It is assumed that the
membrane offers a certain resistance to the current flow for each ion
type, thus preserving the concept that each ion type has its own
specialized channels. It is further assumed that the magnitude of the
current, due to an ionic species, is proportional to the difference

1

between the existing membrane potential and the Nernst potential for that species.

Consider a small, uniformly polarized patch of membrane and suppose that the current through it, due to ion species i, is I_i. Let the Nernst potential be V_i for that type of ion. Then the following equation holds:

$$I_i = g_i(V_i - V),\qquad\qquad(7.2)$$

where V is the membrane potential. Equation (7.2) is taken as the *definition of* g_i, which is called the *chord conductance* of that patch of membrane to ion species i. This heuristic approach enables us to treat membrane currents due to ion fluxes on the same footing as those in the previous cable theory and seems to have been introduced by Hodgkin and Huxley (1952d). The notion of permeability (see Chapter 2) of the cell membrane for an ion species is replaced by that of membrane conductance or resistance for an ion species.

The inclusion of these ionic currents in cable theory is as follows. Let there be n ion species to which the membrane is permeable. Suppose that V_i, assumed constant, is now the Nernst potential for the ith species minus the resting membrane potential. If V is the depolarization, or membrane potential minus the resting potential, then Equation (7.2) still applies. Suppose further that the conductance per unit length at (x, t) for species i is $g_i(x, t)$ (in mhos per centimeter). For reasons, which will become apparent shortly, we introduce a *pump current P* (in amperes per centimeter), which contains ion fluxes due to active transport, heretofore neglected. Then, by considering the membrane current in $(x, x + \Delta x)$ as we did in Chapter 4, Section 2, we obtain the following equation for the depolarization:

$$c_m V_t = \frac{1}{r_i} V_{xx} + \sum_{i=1}^{n} g_i(V_i - V) - P + I_A,\qquad\qquad(7.3)$$

where c_m is the membrane capacitance per unit length (in farads per centimeter), r_i is the internal or axial resistance per unit length (in ohms per centimeter), and I_A is the applied current density (in amperes per centimeter). The usual contributions to the ionic current come from sodium ions, potassium ions, chloride ions with calcium, magnesium, and other cations sometimes being included in special circumstances.

An alternative form of Equation (7.3) and its connection with the previous cable equation

Equation (7.3) is rarely encountered in its present form, though perhaps it should be used in most circumstances. Without the pump term P, it is the basic equation of the Hodgkin–Huxley theory (see Chapter 8). We will obtain another form of (7.3), which will show its connection with the "regular" cable equation of the previous three chapters.

To this end, we separate each ionic conductance g_i into a component $\tilde{g}_i$, constant in space and time, which we associate with the resting membrane, and another component Δg_i, which depends on x and t,

$$g_i(x, t) = \tilde{g}_i + \Delta g_i(x, t). \tag{7.4}$$

Thus Δg_i is the increase in conductance relative to resting level, per unit length, for ion species i. Substituting in (7.3), we obtain

$$c_m V_t = \frac{1}{r_i} V_{xx} + \sum_{i=1}^n \tilde{g}_i V_i - V \sum_{i=1}^n \tilde{g}_i + \sum_{i=1}^n \Delta g_i (V_i - V) - P + I_A. \tag{7.5}$$

Define

$$g_m = \frac{1}{r_m} = \sum_{i=1}^n \tilde{g}_i, \tag{7.6}$$

which is the resting membrane conductance per unit length, where r_m is the resting membrane resistance of unit length times unit length. When there is no applied current, so that $I_A = 0$, and when $\Delta g_i = 0$, so that the ionic conductances have their resting values, there must be zero ionic current. This requires the condition

$$P = \sum_{i=1}^n \tilde{g}_i V_i, \tag{7.7}$$

so that the resting pump rate counters the resting ionic fluxes in order to maintain equilibrium. Then (7.5) can be written as

$$c_m V_t = \frac{1}{r_i} V_{xx} - \frac{V}{r_m} + \sum_{i=1}^n \Delta g_i (V_i - V) + I_A. \tag{7.8}$$

Converting to dimensionless space and time variables, in which the units of space and time are the resting membrane's space and time constants, defined as in Section 4.2, we now have

$$V_t = V_{xx} - V + r_m \sum_{i=1}^n \Delta g_i (V_i - V) + r_m I_A. \tag{7.9}$$

The connection with the previously employed cable equation is as follows. Either one lumps synaptic and other ionic currents into the term I_A and neglects the other forcing term; or one assumes that V_i is so large and changes in V are so relatively small that $\Delta g_i(V_i - V)$ can be regarded as independent of V. The latter may be a valid approximation when considering the sodium ion current, but not usually the potassium and chloride currents.

Equation (7.9) may be rewritten as

$$V_t = V_{xx} - V + \sum_{i=1}^{n} \Delta g_i^*(V_i - V) + r_m I_A, \qquad (7.10)$$

where

$$\Delta g_i^* = \Delta g_i / \sum_{i=1}^{n} \tilde{g}_i \qquad (7.11)$$

is a dimensionless quantity that is the ratio of the increase in the membrane conductance for species i to the total resting membrane conductance.

The boundary conditions for Equations (7.3), (7.8), (7.9) or (7.10) will be the same as those for the previously employed cable equation. Unfortunately, however, the Green's function method of solution does not provide explicit solutions of these equations. Because of the difficulties in solving them, these equations have been neglected in favor of those of the previous cable theory, where exact solutions are more readily available. Some calculations have nevertheless been performed, including those by MacGregor (1968), Barrett and Crill (1974), and Poggio and Torre (1981).

7.2   Equations for the potential with space clamp and synaptic input: reversal potentials
Solution of the space-clamped version of Equation (7.10) with $I_A = 0$

First, consider the space-clamped version of (7.10). That is, set the space derivative (and hence V_{xx}) equal to zero and deal with a small patch of membrane across which the potential is uniform in space. Then, in the absence of applied currents, $V = V(t)$ satisfies the ordinary differential equation

$$\frac{dV}{dt} = -V + \sum_{i=1}^{n} \Delta g_i^*(t)[V_i - V]. \qquad (7.12)$$

During a synaptic event, the ionic conductances will undergo brief increases. These may have different durations for each ion species but we will assume that simultaneous step changes in the conductances occur for each ion type. We let the step changes start at $t = 0$ and end at $t = t_1$,

$$\Delta g_i^*(t) = a_i[H(t) - H(t - t_1)], \qquad t_1 > 0, \qquad (7.13)$$

where $H(\cdot)$ is the unit step function and the $a_i \geq 0$ are constants. Prior to the conductance increase we assume

$$V(0^-) = V_0. \qquad (7.14)$$

To solve (7.12), we note that for $0 \leq t < t_1$, the equation is

$$\frac{dV}{dt} + V\left[1 + \sum_{i=1}^{n} a_i\right] = \sum_{i=1}^{n} a_i V_i. \qquad (7.15)$$

This is a first-order equation in standard form (cf. Section 2.4). We find, on defining

$$A = 1 + \sum_{i=1}^{n} a_i, \qquad (7.16)$$

$$B = \sum_{i=1}^{n} a_i V_i, \qquad (7.17)$$

that, while the conductances are switched on,

$$V(t) = \frac{B}{A} + \left(V_0 - \frac{B}{A}\right) e^{-At}, \qquad 0 \leq t < t_1. \qquad (7.18)$$

If the conductances were switched on indefinitely, V would approach the steady-state value

$$V(\infty) = B/A. \qquad (7.19)$$

The change in $V(t)$ from the beginning to the end of the conductance increases is

$$V(t_1) - V_0 = (B/A - V_0)(1 - e^{-At_1}). \qquad (7.20)$$

Thus whether the change is positive or negative depends on both the initial value V_0 and the quantities A and B. Since A is necessarily positive, we see that synaptic potentials elicited from rest ($V_0 = 0$) will be excitatory if $B > 0$ and inhibitory if $B < 0$. For an excitatory synapse the change in depolarization will be positive if $V_0 < B/A$ and negative if $V_0 > B/A$. Thus the quantity B/A is a "reversal potential" and we have

$$"V_{\text{rev}}" = \frac{\sum_{i=1}^{n} a_i V_i}{1 + \sum_{i=1}^{n} a_i}. \qquad (7.21)$$

This expression is linear in the Nernst potentials for the various ions.

From Chapter 2, Section 14, we have, in general, that V_{rev} is not linear in the individual Nernst potentials, except when only one kind of ion is involved. This deficiency is traced to the inconsistency of (7.2) and the constant-field theory of Chapter 2.

For $t \geq t_1$, when the conductance increases are switched off, the depolarization will be

$$V(t) = (B/A - V_0)(1 - e^{-At_1})e^{-(t-t_1)}. \tag{7.22}$$

Thus, while the synapse is active, the time constant is $1/A$ of its value during the passive decay stage.

An illustrative example

Suppose $V_K = -20$ mV, $V_{Na} = +120$ mV relative to a resting value of -70 mV, and that a chemical transmitter (e.g., acetylcholine) increases the conductances to Na^+ and K^+ only in such a way that $a_{Na} = 0.95$ and $a_K = 0.05$. If the conductance increases last for 0.05 time constants, what is the time course of the potential if the cell is initially at rest? What is the reversal potential?

It will be seen that $A = 2$ and $B = 113$ mV. Hence, from formula (7.18) with $V_0 = 0$, we have

$$V(t) = 56.5(1 - e^{-2t}), \qquad 0 \leq t < 0.05,$$

and, from formula (7.22),

$$V(t) = 5.38e^{-(t-0.05)}, \qquad t \geq 0.05.$$

Thus the peak voltage or EPSP amplitude is 5.38 mV.

From formula (7.21) the reversal potential is

$$V_{rev} = \frac{120 \times 0.95 - 20 \times 0.05}{2}$$

$$= 56.5 \text{ mV}.$$

That is, the reversal potential is -13.5 mV relative to zero membrane potential.

Note that, as is sometimes possibly the case, if only one ion species is involved, then the corresponding a_i can be calculated from the reversal potential if the Nernst potential is known.

Response to synaptic input with injected current

In certain experiments, such as those mentioned in Section 2.14, a constant current I_0, which may be hyperpolarizing or depolarizing, is injected into a nerve cell; and then, usually after a steady state has been achieved, synaptic inputs are activated. For this situation in the space-clamped case we have, with synaptic-input conductance changes lasting from $t = 0$ to $t = t_1$,

$$\frac{dV}{dt} + V = \sum_{i=1}^{n} a_i [H(t) - H(t - t_1)][V_i - V] + V_0, \qquad t > 0,$$

$$(7.23)$$

where, now,

$$V_0 = r_m I_0 \qquad (7.24)$$

is the steady-state voltage under the application of the current I_0.

If the steady state has been achieved by the time the synaptic input occurs, then we must solve (7.23) with the initial condition

$$V(0) = V_0. \qquad (7.25)$$

It is left as an exercise to show that the time course of the potential, while the synapses are active, is given by

$$V(t) = V_0 + \tilde{B}(1 - e^{-At})/A, \qquad 0 \leq t \leq t_1, \qquad (7.26)$$

where

$$\tilde{B} = \sum_{i=1}^{n} a_i (V_i - V_0). \qquad (7.27)$$

The resulting change in potential is thus positive if $\tilde{B} > 0$ and negative if $\tilde{B} < 0$. Equivalently, the response is depolarizing (excitatory) if $V_0 < \sum a_i V_i / \sum a_i$ and hyperpolarizing (inhibitory) if $V_0 > \sum a_i V_i / \sum a_i$. Hence the reversal potential is

$$V_{\text{rev}} = \left(\sum_{i=1}^{n} a_i V_i \right) \Big/ \left(\sum_{i=1}^{n} a_i \right). \qquad (7.28)$$

If only sodium and potassium ions are involved, this formula can be written as

$$V_{\text{rev}} = \frac{(a_{\text{Na}}/a_{\text{K}})V_{\text{Na}} + V_{\text{K}}}{(a_{\text{Na}}/a_{\text{K}}) + 1}, \qquad (7.29)$$

which is given by Kuffler and Nicholls (1976), page 167.

EPSPs obtained from Equation (7.26) are illustrated in Figure 7.1 for various values of the injected current, which takes the cell to various steady depolarized states. Note that the time constant is smaller while the synaptic input is on–the decay phase has the "membrane time constant."

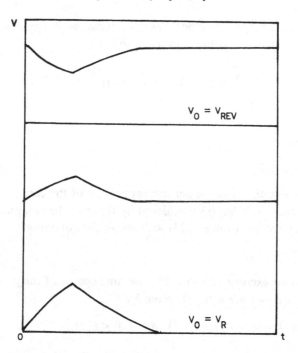

Figure 7.1. Reversal potential for excitatory synaptic input. As the steady depolarization increases, the EPSP amplitude decreases, becomes zero at the reversal potential, and thereafter exhibits a hyperpolarizing appearance.

7.3 **Cable equation with excitatory and inhibitory reversal potentials**

We will derive a version of (7.9), valid under certain assumptions, which is particularly useful in delineating excitatory and inhibitory inputs. First, we distinguish those ionic conductance changes induced by excitatory synaptic input from those due to inhibitory input. If n_E ion species are involved in excitation and n_I in inhibition, then

$$
V_t = -V + V_{xx} + r_m \left[\sum_{i=1}^{n_E} \Delta g_{i,E}(V_{i,E} - V) \right.
$$
$$
\left. + \sum_{i=1}^{n_I} \Delta g_{i,I}(V_{i,I} - V) \right] + r_m I_A
$$
$$
= -V + V_{xx} + r_m \left[\sum_{i=1}^{n_E} \Delta g_{i,E} \left\{ \frac{\Sigma \Delta g_{i,E} V_{i,E}}{\Sigma \Delta g_{i,E}} - V \right\} \right.
$$
$$
\left. + \sum_{i=1}^{n_I} \Delta g_{i,I} \left\{ \frac{\Sigma \Delta g_{i,I} V_{i,I}}{\Sigma \Delta g_{i,I}} - V \right\} \right] + r_m I_A.
$$
$$
(7.30)
$$

If there is only one ion species involved in excitation and only one involved in inhibition, then the Nernst potentials $V_{1,E}$ and $V_{1,I}$ will be the synaptic reversal potentials, which we denote by V_E and V_I. Then (7.30) becomes

$$V_t = -V + V_{xx} + r_m[\Delta g_E(V_E - V) + \Delta g_I(V_I - V)] + r_m I_A,$$
(7.31)

where $\Delta g_E(x, t) = \Delta g_{1,E}(x, t)$ and $\Delta g_I(x, t) = \Delta g_{1,I}(x, t)$.

We will show that (7.31) is sometimes applicable in situations where there are several ion species involved in the synaptic transmission. Consider a single excitatory synapse extending from x_E to x'_E, which is activated from t_E to t'_E where the (assumed) constant magnitudes of the conductance increases for the i, E ion species are $a_{i,E}$. Then, abbreviating $H(x - x_E)$ to $H(x_E)$, etc.,

$$\Delta g_{i,E}(x, t) = a_{i,E}[H(x_E) - H(x'_E)][H(t_E) - H(t'_E)],$$
$$i = 1, \ldots, n_E.   (7.32)$$

Assume a similar set of conductance changes for an inhibitory synapse. Then, substituting in (7.30) gives

$$V_t = -V + V_{xx} + r_m\left[\{H(x_E) - H(x'_E)\}\{H(t_E) - H(t'_E)\}\left(\sum_{i=1}^{n_E} a_{i,E}\right)\right.$$
$$\times\left\{\frac{\sum a_{i,E}V_{i,E}}{\sum a_{i,E}} - V\right\} + \{H(x_I) - H(x'_I)\}$$
$$\times\{H(t_I) - H(t'_I)\}\left(\sum_{i=1}^{n_I} a_{i,I}\right)$$
$$\left.\times\left\{\frac{\sum a_{i,I}V_{i,I}}{\sum a_{i,I}} - V\right\}\right] + r_m I_A.   (7.33)$$

But

$$\frac{\sum_i a_{i,E}V_{i,E}}{\sum_i a_{i,E}} = V_E,   (7.34)$$

and

$$\frac{\sum_i a_{i,I}V_{i,I}}{\sum_i a_{i,I}} = V_I   (7.35)$$

are, according to (7.28), the excitatory and inhibitory reversal potentials. If we set

$$\Delta g_E = \sum_{i=1}^{n_E} a_{i,E},   (7.36A)$$

$$\Delta g_I = \sum_{i=1}^{n_I} a_{i,I},   (7.36B)$$

then (7.33) becomes identical to (7.31). Furthermore, if there are several synapses of either excitatory or inhibitory character, and these are active on several different time intervals, then, providing the corresponding conductance changes are of the kind in Equation (7.32), Equation (7.31) will still apply (additivity). This represents a simplification even though (7.31) is no easier to solve than the more general equation (7.30). Although something is lost in going from (7.30) to (7.31), we will employ (7.31) in the remainder of this chapter because it is more widely used and makes more transparent the contributions from excitation and inhibition. Any solutions obtained from Equation (7.31) can, of course, be employed for (7.30) with a change of symbols.

When the radius of the dendrite (or possibly axon) is not constant but is a function $a(x)$ of the (physical) distance x from the origin, the cable equation is modified as in Section 6.6. Although we will not be concerned with its solutions, the cable equation including reversal potentials and a varying radius is

$$\tau V_t = -V + \lambda_0^2/F(x)\left[\left(R^2(x)V_x\right)_x + I(x,t)\right]$$
$$+ \rho_m \delta\left[g_E(x,t)(V_E - V) + g_I(x,t)(V_I - V)\right], \quad (7.37)$$

where $g_E(x,t)$ and $g_I(x,t)$ are the excitatory and inhibitory conductances per unit area (in $(x, x + \Delta x])$; the remaining symbols were explained in Section 6.6.

7.4 Space-clamped equation with reversal potentials
7.4.1 Effects of timing of excitation and inhibition

The space-clamped version of Equation (7.31) in the absence of externally applied currents is

$$dV/dt + V = r_m\left[\Delta g_E(V_E - V) + \Delta g_I(V_I - V)\right]. \quad (7.38)$$

The variety of patterns of arrival of excitation and inhibition is extremely large and we will examine here only the case of single excitatory and inhibitory inputs arriving at different times and having various durations. Let the excitatory synapses be active from 0 to t_E and the inhibitory synapses be active from t_I to t_I'. Assuming the conductances are constant while the inputs are active, we set

$$r_m \Delta g_E = a_E\left[H(t) - H(t - t_E)\right], \quad (7.39)$$
$$r_m \Delta g_I = a_I\left[H(t - t_I) - H(t - t_I')\right]. \quad (7.40)$$

We examine the response of a cell that is initially at rest

$$V(0) = 0. \quad (7.41)$$

The excitation and inhibition may overlap in time or they may not. The latter case is treated first.

(*a*) *No overlap of excitation and inhibition*
If the intervals $(0, t_E)$ and (t_I, t_I') do not overlap, we have

$$V(t) = A_E[1 - \exp\{-\tau_E t\}], \qquad 0 \le t \le t_E, \qquad (7.42)$$

where

$$A_E = a_E V_E/(1 + a_E) \qquad (7.43)$$

is the asymptotic voltage if the excitation is on indefinitely, and

$$\tau_E = 1 + a_E \qquad (7.44)$$

is the effective time constant. After the excitation and prior to the inhibition,

$$V(t) = V(t_E)\exp\{-(t - t_E)\}, \qquad t_E \le t \le t_I. \qquad (7.45)$$

Then

$$V(t) = V(t_I)\exp\{-\tau_I(t - t_I)\}$$
$$+ A_I[1 - \exp\{-\tau_I(t - t_I)\}], \qquad t_I \le t \le t_I', \quad (7.46)$$

where

$$A_I = a_I V_I/(1 + a_I), \qquad (7.47)$$
$$\tau_I = 1 + a_I. \qquad (7.48)$$

After t_I', the voltage again decays exponentially

$$V(t) = V(t_I')\exp\{-(t - t_I')\}, \qquad t \ge t_I'. \qquad (7.49)$$

The response will be as sketched in Figure 7.2.

The effect of timing may be seen by determining the magnitude of the voltage change $V(t_I) - V(t_I')$ during the inhibition as a function of $t_I - t_E$ and comparing it with the voltage change when there is infinite separation between the excitation and inhibition. Denoting the voltage change during the inhibition by $\Delta V(t_I - t_E)$, we find, from (7.42)–(7.47),

$$\frac{\Delta V(t_I - t_E)}{\Delta V(\infty)} = 1 - \left(\frac{\Delta V_{ex}}{A_I}\right)\exp\{-(t_I - t_E)\}, \qquad (7.50)$$

where

$$\Delta V_{ex} = A_E[1 - \exp\{-\tau_E t_E\}] \qquad (7.51)$$

is the voltage change during the excitation, and

$$\Delta V(\infty) = -A_I[1 - \exp\{-\tau_I(t_I' - t_I)\}]. \qquad (7.52)$$

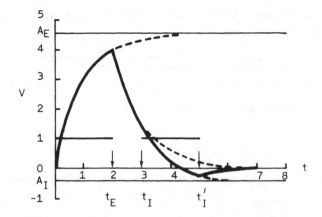

Figure 7.2. Time course of the voltage with excitation from 0 to t_E, followed by inhibition from t_I to t_I'. For an explanation of the symbols, see the text. The dashed lines indicate projected values if subsequent conductance changes do not occur.

The ratio (7.50) is maximal at unity when $t_I - t_E = \infty$, $\Delta V_{ex} = 0$, or $A_I = \infty$. Excluding the latter two cases, the effect of the inhibition is greater the longer the time interval between the excitation and inhibition and it is exponential in time. Stronger inhibitory inputs are less affected by prior excitation; also, the smaller the prior excitation, the smaller the disinhibitory effect of the excitation – the exact relation that is given by (7.50).

(b) Overlap of excitation and inhibition

Assume again that the excitation arrives first. Then there are two possibilities. The excitation may be over before the inhibition ceases ($t_E < t_I'$) or vice versa. For the first case, $0 < t_I < t_E < t_I' < \infty$.

For $0 \le t \le t_I$, while only the excitation is on, formula (7.42) applies. Thereupon

$$V(t) = V(t_I) \exp\{-\tau_{EI}(t - t_I)\}$$
$$+ A_{EI}[1 - \exp\{-\tau_{EI}(t - t_I)\}], \qquad t_I \le t \le t_E, \tag{7.53}$$

where

$$\tau_{EI} = 1 + a_E + a_I, \tag{7.54}$$

$$A_{EI} = \frac{a_E V_E + a_I V_I}{1 + a_E + a_I}. \tag{7.55}$$

For $t_E \le t \le t_I'$, the voltage is given by (7.46) with t_I replaced by t_E.

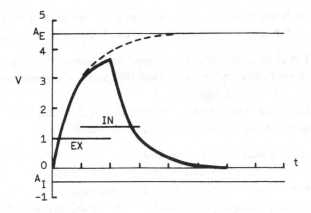

Figure 7.3. Time course of the voltage when the inhibitory and excitatory inputs overlap in time. Dashed lines indicate projected voltages if subsequent conductance changes do not occur.

Subsequently, the voltage decays exponentially with time constant unity. An example is illustrated in Figure 7.3.

In this case the voltage change induced by the inhibitory input depends on the lengths of the three time intervals $(0, t_E)$, (t_I, t_E), and (t_E, t_I'). Comparing the voltage change with that which occurs when there is infinite separation gives

$$
\frac{\Delta V}{\Delta V(\infty)} = \left[[1 - \exp\{-\tau_I(t_I' - t_E)\}] + \left(\frac{A_{EI}}{A_I} \right) \right.
$$
$$
\times [1 - \exp\{-\tau_{EI}(t_E - t_I)\}] \exp\{-\tau_I(t_I' - t_E)\}
$$
$$
- \left(\frac{A_E}{A_I} \right) [1 - \exp\{-\tau_E t_I\}]
$$
$$
\left. \times [1 - \exp\{-\tau_{EI}(t_E - t_I)\} \exp\{-\tau_I(t_I' - t_E)\}] \right] \Big/
$$
$$
[1 - \exp\{-\tau_I(t_I' - t_E) - \tau_I(t_E - t_I)\}]. \quad (7.56)
$$

Recurrent inhibition

In Chapter 1 it was pointed out that motoneuron axon collaterals form excitatory synapses with Renshaw cells, which exert negative feedback on the motoneurons through inhibitory synapses. The purpose of this recurrent inhibition has been postulated as twofold. First, it exerts a "stabilizing" effect on motoneuron discharge, since the greater the firing rate of the motoneurons, ignoring the complications of depression and facilitation, the greater the inhibitory input from Renshaw cells. The second functional role of the recurrent inhibition is more involved. It has been established that:

(a) Renshaw cells receive stronger (excitatory) input from the larger phasic motoneurons than the smaller tonic motoneurons (cf. Chapter 1); and

(b) the tonic motoneurons receive stronger (inhibitory) input from the Renshaw cells than do the phasic cells.

It has been concluded from (a) and (b) that an inhibitory effect is aimed at the tonic motoneurons by the phasic cells and that this inhibition is mediated by Renshaw cells. Thus, when rapid movements such as running, jumping, and so on, are executed (phasic cells), the discharge from motoneurons involved in steady and sustained postures (tonic cells) is silenced to ensure noninterference. For further details, see Eccles et al. (1961), Eccles (1964), Granit (1963, 1970), Curtis and Ryall (1966), Pompeiano and Wand (1976), and Eccles (1977).

The timing of the inhibition from Renshaw cells to motoneurons is clearly important. When a strong volley of impulses is set up in several motoneurons, Renshaw cells fire a vigorous burst of action potentials with an initial frequency as high as 1000 to 1500 s^{-1}. The arrival of the inhibitory inputs at the motoneurons is only about 1 ms after the motoneuron spikes (Eccles, 1977).

7.4.2 Repetitive excitatory synaptic input

Consider the case in which excitatory synapses are activated with a constant frequency $f_{in} = 1/T$. During each synaptic input a conductance increase occurs and the equation for the subthreshold depolarization is

$$\frac{dV}{dt} + V = a_E[V_E - V] \sum_{n=0}^{\infty} [H(t-nT) - H(t-(nT+t_E))],$$

$$V(0) = 0, \quad (7.57)$$

where the duration of the conductance change is t_E. It is an exercise to show that, assuming no overlap of inputs ($t_E \le T$),

$$V(nT)$$

$$= \frac{A_E[1 - \exp(-\tau_E t_E)] \exp(-(T-t_E))[1 - \exp(-n(T+a_E t_E))]}{1 - \exp(-(T+a_E t_E))},$$

$$n = 0,1,\dots. \quad (7.58)$$

If a steady state is achieved, the minima of V will be

$$V_{\min} = \frac{A_E[1 - \exp(-\tau_E t_E)]\exp(-(T - t_E))}{1 - \exp(-(T + a_E t_E))},\tag{7.59}$$

and the maxima will be

$$V_{\max} = \frac{A_E[1 - \exp(-\tau_E t_E)]}{1 - \exp(-(T + a_E t_E))}.\tag{7.60}$$

If a constant threshold θ for firing is assumed, the condition for firing is

$$V_{\max} \geq \theta.\tag{7.61}$$

Frequency of firing

Given that (7.61) is satisfied, what will be the frequency of firing? Assuming an initial resting state, it may be the case that one synaptic input is sufficient to take V to threshold. In this case,

$$A_E[1 - \exp(-\tau_E t_E)] \geq \theta,\tag{7.62}$$

and the time to reach threshold is

$$T_\theta = -\frac{1}{\tau_E}\ln\left[1 - \frac{\theta}{A_E}\right],\tag{7.63}$$

independently of the input frequency. The frequency of firing is then equal to tne input frequency.

When more than one synaptic input is required to take the cell to threshold, T_θ can be found as follows. The depolarization at the *end* of the nth input is

$$V(nT + t_E) = V(nT)\exp(-\tau_E t_E) + A_E[1 - \exp(-\tau_E t_E)],\tag{7.64}$$

where $V(nT)$ is given in (7.58). Now, since $V(nT + t_E)$ is an increasing function of n, regard n as continuous and solve

$$V(xT + t_E) = \theta,\tag{7.65}$$

for x. If x is an integer, then

$$T_\theta = xT + t_E.\tag{7.66}$$

If x is not an integer, then let

$$n_x = \text{largest integer in } [0, x), \qquad x > 0.\tag{7.67}$$

Then $n_x + 2$ inputs are required for V to reach θ. (Note that the first input is at $t = 0$.) Suppose the $(n_x + 2)$th input must be active for a duration of T_1 before θ is reached. Then

$$T_\theta = (n_x + 1)T + T_1,\tag{7.68}$$

which holds for all $x > 0$. An absolute refractory period may be added in (7.68).

Explicitly, we have, from (7.58) and the solution of (7.57),

$$x = \frac{1}{T + a_E t_E} \ln\left[\frac{1}{1-Y}\right], \tag{7.69}$$

$$Y = \left(1 - \frac{\theta}{A_E[1 - \exp(-\tau_E t_E)]}\right)(1 - \exp(T + a_E t_E)), \tag{7.70}$$

$$T_1 = \frac{1}{\tau_E}\ln\left[\frac{A_E - V((n_{x+1})T)}{A_E - \theta}\right]. \tag{7.71}$$

Note that condition (7.61) implies $A_E > \theta$ when $T > t_E$. The argument of the logarithm in (7.69) is therefore positive since $V((n_x + 1)T) < \theta$. Figure 7.4 shows some computed input–output-frequency curves assuming a refractory period of 0.2.

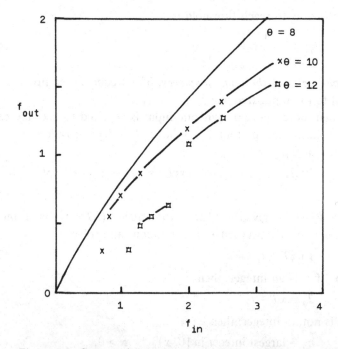

Figure 7.4. Computed input–output-frequency relations for repetitive synaptic excitation using the space-clamped model with reversal potential. Parameter values: $a_E = 1$, $t_E = 0.2$, and $V_E = 50$. Units of f are reciprocal time constants. For $\theta = 8$ only one input is required for V to reach θ. Curves for $\theta = 10, 12$ have discontinuities as 2, 3, etc., inputs are necessary for firing.

7.5 Response to impulsive conductance changes: nonlinear interaction

Although synaptic conductance changes are sustained in time, if they are short lasting it is expeditious to consider them as instantaneous (MacGregor, 1968).

7.5.1 A periodic train of excitation

We consider repetitive excitatory conductance pulses occurring at the point $x = x_E$ and separated by T time units. The equation for V is

$$V_t = V_{xx} - V + a_E \delta(x - x_E)(V_E - V) \sum_{n=0}^{\infty} \delta(t - nT), \quad (7.72)$$

where a_E is a constant and V_E is the excitatory reversal potential. We let the interval of values of x be arbitrary for now but assume that we have boundary conditions at the ends of the interval such that a Green's function can be found for the usual cable equation $V_t = V_{xx} - V$. Let $G(x, y; t)$ be this Green's function. The Green's function formula will apply in the following sense. In general, the solution of

$$V_t = V_{xx} - V + g(x, t)(V_E - V), \quad (7.73)$$

with initial condition

$$V(x, 0) = v(x), \quad (7.74)$$

is

$$V(x, t) = \int G(x, y; t) v(y) \, dy + \iint [V_E - V(y, s)]$$

$$\times g(y, s) G(x, y; t - s) \, ds \, dy. \quad (7.75)$$

That is, the source term is $(V_E - V)g$, differing from the cases met previously in that it involves V, the function that is sought.

If an impulse is delivered at $t = 0$ and the potential is initially zero, then

$$V(x, t) = a_E V_E G(x, x_E; t), \quad 0 < t \le T. \quad (7.76)$$

This is the response to a single impulse and does not differ in form from that of the neuron without synaptic reversal potentials.

Now we regard

$$V(x, T) = a_E V_E G(x, x_E; T), \quad (7.77)$$

as the initial condition when the next impulsive conductance change occurs. Let $t' = t - T$ so t' measures the time elapsed since the second synaptic input. Applying formula (7.75) again, we have

$$V(x, t') = \int a_E V_E G(y, x_E; T) G(x, y; t') \, dy$$

$$+ \iint a_E [V_E - V(y, s)] \, \delta(y - x_E) \, \delta(s)$$

$$\times G(x, y; t' - s) \, ds \, dy. \tag{7.78}$$

The first integral may be evaluated by using the (semigroup) property of the Green's function

$$\int G(x, y; s) G(z, x; t) \, dx = G(z, y; s + t), \tag{7.79}$$

whose proof is left as an exercise.

Evaluating the double integral gives

$$a_E [V_E - V(x_E, t' = 0)] G(x, x_E; t')$$

$$= a_E [V_E - V(x_E, T)] G(x, x_E; t').$$

Using (7.77), we get

$$V(x, t) = a_E V_E G(x, x_E; t) + a_E V_E G(x, x_E; t - T)$$

$$\times [1 - a_E G(x_E, x_E; T)], \quad T < t \le 2T. \tag{7.80}$$

In particular,

$$V(x, 2T) = a_E V_E [1 - a_E G(x_E, x_E; T)] G(x, x_E; T)$$

$$+ a_E V_E G(x, x_E; 2T). \tag{7.81}$$

Proceeding to the next time interval, $2T < t \le 3T$, we find in the same way that $V(x, 3T)$ consists of a linear combination of $G(x, x_E; T)$, $G(x, x_E; 2T)$, and $G(x, x_E; 3T)$. We infer that, in general, $V(x, nT)$ can be expressed as linear combination of $G(x, x_E; T), \ldots,$ $G(x, x_E; nT)$.

A recursive formula for $V(x, nT)$
Given this clue, let us set

$$V(x, nT) \doteq \sum_{k=1}^{n} \alpha_{nk} G(x, x_E; kT), \tag{7.82}$$

where the constant coefficients α_{nk} are found as follows. For $nT < t$

$\leq (n+1)T$, put $t' = t - nT$. Then, in this time interval,

$$V(x, t') = \int \sum_{k=1}^{n} \alpha_{nk} G(y, x_E; kT) G(x, y; t') \, dy$$

$$+ \iint G(x, y; t' - s) a_E [V_E - V(y, s)]$$

$$\times \delta(y - x_E) \delta(s) \, ds \, dy$$

$$= \sum_{k=1}^{n} \alpha_{nk} G(x, x_E; t' + kT) + a_E G(x, x_E; t')$$

$$\times [V_E - V(x_E, t' = 0)], \tag{7.83}$$

or

$$V(x, t) = \sum_{k=1}^{n} \alpha_{nk} G(x, x_E; t - nT + kT)$$

$$+ a_E G(x, x_E; t - nT) \left[V_E - \sum_{k=1}^{n} \alpha_{nk} G(x_E, x_E; kT) \right],$$

$$nT < t \leq (n+1)T. \tag{7.84}$$

Substituting $t = (n+1)T$ in this formula, we have

$$V(x, (n+1)T) = \sum_{k=1}^{n} \alpha_{nk} G(x, x_E; (k+1)T)$$

$$+ a_E G(x, x_E; T) \left[V_E - \sum_{k=1}^{n} \alpha_{nk} G(x_E, x_E; kT) \right]$$

$$= a_E \left[V_E - \sum_{k=1}^{n} \alpha_{nk} G(x_E, x_E; kT) \right] G(x, x_E; T)$$

$$+ \sum_{k=1}^{n} \alpha_{nk} G(x, x_E; (k+1)T)$$

$$\doteq \sum_{j=1}^{n+1} \alpha_{n+1, j} G(x, x_E; jT). \tag{7.85}$$

We can now read off the recursive relation, which gives, in summary,

$$V(x, nT) = \sum_{k=1}^{n} \alpha_{nk} G(x, x_E; kT), \tag{7.86}$$

$$\alpha_{11} = a_E V_E, \tag{7.87}$$

$$\alpha_{n+1, 1} = a_E \left[V_E - \sum_{k=1}^{n} \alpha_{nk} G(x_E, x_E; kT) \right], \tag{7.88}$$

$$\alpha_{n+1, k} = \alpha_{n, k-1}, \qquad k = 2, 3, \ldots, n+1. \tag{7.89}$$

The full time-dependent solution is given by (7.84) when the coefficients α_{nk} are found from (7.87)–(7.89).

Assuming that the cell is capable of producing spikes and given a suitable threshold condition and boundary conditions, the time interval between output spikes can be computed from the above formulas. Such calculations cannot be done analytically and are quite lengthy, especially for finite cables. MacGregor (1968) has performed such calculations for excitatory inputs at various locations for an infinite cable with an assumed exponentially decaying threshold at the soma ($x = 0$). An interesting claim was that the frequency-transfer curves saturate at high-input frequencies due to the diminution of the amplitude of the postsynaptic potentials.

The steady state

Experimental evidence [e.g., Curtis and Eccles (1960); also see Figure 3.10] points to the existence of a steady-state distribution of potential when synaptic inputs are stimulated indefinitely, provided conditions for action-potential instigation are not achieved. Examination of the recursion relations (7.88) and (7.89) shows that $\alpha_{\infty k} = \alpha$, $k = 1, 2, \ldots$. Hence, from (7.88),

$$\alpha = \frac{a_E V_E}{1 + a_E \sum_{k=1}^{\infty} G(x_E, x_E; kT)}, \qquad (7.90)$$

and the steady-state distribution of potential at the time of arrival of the input is

$$V^*(x) = \alpha \sum_{k=1}^{\infty} G(x, x_E; kT). \qquad (7.91)$$

The infinite series in (7.91) converges; at least this is readily verifiable for infinite cables where their terms are bounded by those of a geometric series.

Whereas formulas (7.90) and (7.91) enable the steady-state voltage to be found at the times of arrival of the synaptic input, in order to find the steady-state distribution of potential *between* inputs, $V^*(x)$ must be used as the initial condition in conjunction with an impulsive conductance change at "$t = 0$" (quotes indicate that reference is to infinite times). From formula (7.75), the steady-state EPSP is given by

$$\tilde{V}(x, t) = \int V^*(y) G(x, y; t)\, dy + a_E \int \int_0^t [V_E - \tilde{V}(y, s)]$$

$$\times \delta(s)\, \delta(y - x_E) G(x, y; t - s)\, ds\, dy, \qquad (7.92)$$

for "$0 < t \leq T$" *in the steady state.* On substituting for V^* from (7.91) and using (7.79), there results

$$\tilde{V}(x,t) = \alpha \sum_{k=1}^{\infty} G(x, x_E; t + kT) + a_E [V_E - V^*(x_E)] G(x, x_E; t).$$

$$(7.93)$$

If this is truly the steady-state potential, it must return at "$t = T$" to its value at "$t = 0$." It will be verified, with the aid of (7.90), that this is indeed the case; namely,

$$\tilde{V}(x, T) = V^*(x). \qquad (7.94)$$

The results obtained for the steady-state EPSP under repetitive impulsive conductance changes should correspond, in a first approximation, to the experimental results on repetitive monosynaptic excitation of various nerve cells reported by Eccles (1969), page 84 et seq. Terminal responses were measured for various frequencies of stimulation, corresponding to different and approximately known values of T. It is possible that on evaluation of the formulas for $V^*(0)$ and $V(0, t)$, "$0 < t \leq T$," which give the voltage at the origin (soma), the parameters a_E, V_E and x_E could be estimated from the experimental data, thus providing information on the nature and location of the synaptic transmission.

Figure 7.5 shows terminal or steady-state EPSPs calculated for the case of repetitive impulsive conductance changes using formula (7.93).

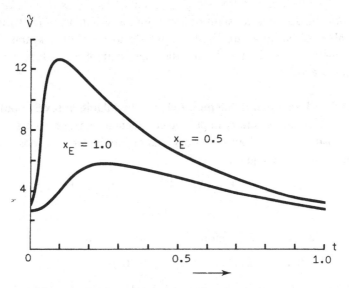

Figure 7.5. Steady-state EPSPs calculated from (7.93) for two input locations. For parameter values, see the text.

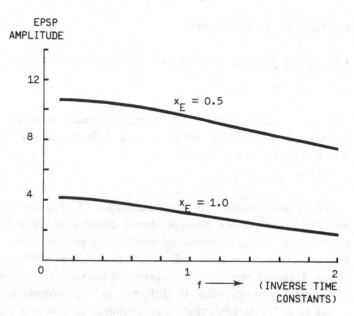

Figure 7.6. Steady-state EPSP amplitude plotted against input frequency (in inverse time constants) for two input locations. For parameter values, see the text.

The Green's function employed was that for infinite cables and the calculated voltage is at $x = 0$. The parameter values employed were $a_E = 0.5$, $V_E = 50$, $T = 1$, and two values of x_E, namely, 0.5 and 1.0. For $T = 1$ the input frequency is about 200 s^{-1} based on a (typical motoneuron) time constant of 5 ms. Similar calculations for various T enable the steady-state EPSP amplitude to be plotted against input frequency as in Figure 7.6 for the same pair of x_E values. See also MacGregor (1968).

7.5.2   Excitation (inhibition) followed by inhibition (excitation)

If an excitatory impulsive conductance change occurs at $t = 0$ at position x_E and an inhibitory one occurs at $t = t_I$ at position x_I, the appropriate equation for V is

$$V_t = V_{xx} - V + a_E[V_E - V]\,\delta(x - x_E)\,\delta(t)$$
$$+ a_I[V_I - V]\,\delta(x - x_I)\,\delta(t - t_I). \tag{7.95}$$

We have, as in (7.76), before the inhibition arrives,

$$V(x, t) = a_E V_E G(x, x_E; t), \qquad 0 < t \le t_I. \tag{7.96}$$

Then $V(x, t_I) = a_E V_E G(x, x_E; t_I)$ is used as an initial condition for an inhibitory pulse at $t = t_I$. Thus, with $v(x) = V(x, t_I)$ and putting

$t' = t - t_I$, we have, from (7.75),

$$V(x, t') = \int G(x, y; t') v(y) \, dy$$

$$+ a_I \int \int_0^{t'} [V_I - V(y, s)] \, \delta(y - x_I) \, \delta(s)$$

$$\times G(x, y; t' - s) \, ds \, dy. \qquad (7.97)$$

Evaluating the double integral, we obtain

$$V(x, t') = a_E V_E \int G(y, x_E; t_I) G(x, y; t') \, dy$$

$$+ a_I [V_I - V(x_I, t_I)] G(x, x_I; t'). \qquad (7.98)$$

The remaining integral is evaluated with the aid of (7.79) to give

$$V(x, t) = a_E V_E G(x, x_E; t) + a_I V_I G(x, x_I; t - t_I)$$

$$- a_E a_I V_E G(x_I, x_E; t_I) G(x, x_I; t - t_I), \qquad t > t_I. \qquad (7.99)$$

It can be seen that the response is the sum of the responses to the individual excitatory and inhibitory synaptic inputs, less an interference term, which is proportional to $a_E a_I V_E$ and goes to zero as the time between the inputs gets larger. Thus the response is not the sum of the individual responses, a phenomenon often referred to as *nonlinear summation*. In words,

$$\text{response} = \text{EPSP} + \text{IPSP} - \text{interference term}. \qquad (7.100)$$

This contrasts with the additivity of responses for the usual cable equation [cf. Equation (5.45)]. It is interesting that the magnitude of the interference term does not depend on V_I, the inhibitory reversal potential.

In addition to enabling one to study the effects of timing of the excitatory and inhibitory pulses as was done in Section 7.4.1 for the space-clamped equation, formula (7.99) may also be used to ascertain quantitatively the effects of the relative positions of the excitatory and inhibitory inputs. To illustrate, the time courses of the voltage were calculated from (7.99) for a fixed excitatory input position and a variable inhibitory input location. The results are shown in Figures 7.7A and 7.7B. In these calculations the infinite cable Green's functions were again employed and the parameter values were $a_E = 0.1$, $a_I = 0.2$, $V_E = 50$, $V_I = -10$, and $t_I = 1.0$. In Figure 7.7A the response at the origin is shown when the excitation is at $x_E = 0.5$ and the inhibition is more remote at $x_I = 0.75$. It can be seen that the effect of the inhibition, though fairly abrupt, is small in its depression of the voltage. On the other hand, when the inhibition is closer to the origin

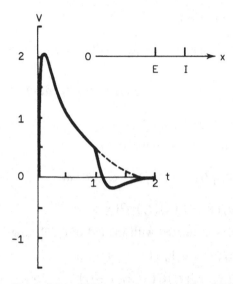

Figure 7.7A. Voltage at the origin when inhibition follows excitation with the inhibition more remote than the excitation. Calculated from (7.99) with parameter values as given in the text.

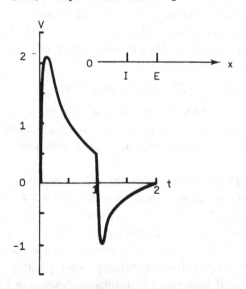

Figure 7.7B. As in Figure 7.7A except that the inhibition is closer to the origin than the excitation.

at $x_I = 0.25$ and the excitation is still at $x_E = 0.5$, the inhibitory input acts rapidly to depress the voltage by a large amount.

It is useful to rewrite Equation (7.99) as

$$V(x,t) = a_E V_E G(x, x_E; t) + a_I V_I G(x, x_I; t - t_I)$$
$$\times [1 - a_E V_E G(x_I, x_E; t_I)/V_I], \qquad t > t_I.$$

Since in most cases $V_I < 0 < V_E$, we see that

$$1 - a_E V_E G(x_I, x_E; t_I)/V_I > 1.$$

Hence the effect of an inhibitory synaptic input is always enhanced by a previous excitatory synaptic input.

If the inhibitory input arrives before the excitatory one, the latter occurring at $t = t_E$, the response will be

$$V(x, t) = a_I V_I G(x, x_I; t) + a_E V_E G(x, x_E; t - t_E)$$
$$\times [1 - a_I V_I G(x_E, x_I; t_E)/V_E], \qquad t > t_E.$$

Thus the previous inhibitory input leads to an amplification of the response to an excitatory input.

In order to quantify the amplification of the response to the second input due to the previous input, we define *amplification factors*. These are the ratio of the response to the second input in the presence of a previous input to the response to the second input alone. For excitation followed by inhibition, the amplification factor is

$$A_{E, I} = 1 + |a_E V_E/V_I| \, G(x_I, x_E; t_I),$$

this formula being valid for several kinds of boundary conditions.

Note that $A_{E, I}$ is an increasing function of a_E, so that the larger the excitatory input, the greater the amplification of the subsequent inhibitory response. $A_{E, I}$ is also a decreasing function of $|x_E - x_I|$ and must achieve a maximum when $x_E = x_I$. That is, the amplification is greater the smaller the distance between the synapses. Also, we must have

$$A_{E, I} \xrightarrow[|x_E - x_I| \to \infty]{} 1,$$

so that as the distance between the synapses becomes larger and larger, the interference between the two responses becomes negligible. With regard to timing, there is a time interval between the inputs at which, for given values of the remaining input parameters, the amplification factor $A_{E, I}$ attains a maximum. This is because the Green's function $G(x_E, x_I; t_I)$, for $x_E \neq x_I$, rises to a maximum as t_I increases and then decays monotonically to zero. Hence, as $t_I \to \infty$, the factor $A_{E, I}$ also tends to unity.

Insight can be obtained quickly by employing, as an approximation, the infinite cable Green's function. For this case the amplification factor is

$$A_{E, I} = 1 + \left| \frac{a_E V_E}{V_I} \right| \frac{e^{-t_I} e^{-(x_I - x_E)^2/4t_I}}{\sqrt{4\pi t_I}}.$$

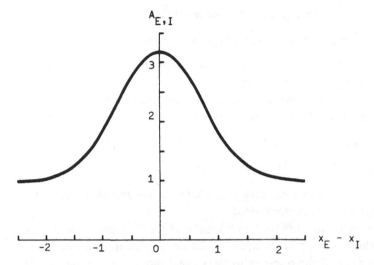

Figure 7.8. Amplification factor for inhibitory input in the presence of a previous excitatory input, as a function of the distance between the excitatory and inhibitory synapses. Parameter values: $a_E = 1$, $V_E = 50$, $V_I = -10$, and $t_I = 0.25$.

Hence the amplification factor, as a function of the distance between the synapses, has the form of unity plus a normal density function. For fixed values of the remaining parameters, $A_{E,I}$ will therefore appear as in Figure 7.8.

Similarly, for inhibition followed by excitation, the amplification of the excitatory response due to the prior inhibitory input is

$$A_{I,E} = 1 + |a_I V_I / V_E| \, G(x_E, x_I; t_E).$$

Shunting inhibition

Koch, Poggio, and Torre (1983) call inhibitory inputs in which the inhibitory reversal potential is approximately equal to the resting potential (i.e., $V_I = 0$) *shunting inhibition*. If we put $V_I = 0$ in Equation (7.99), when an excitatory input is followed by an inhibitory one, we obtain

$$V(x, t) = a_E V_E G(x, x_E; t) - a_E a_I V_E G(x_I, x_E; t_I)$$
$$\times G(x, x_I; t - t_I), \qquad t > t_I.$$

Due, in particular, to the usually large magnitude of V_E, the inhibitory effect of the second pulse can be large even though in the absence of the prior excitatory input the inhibitory synapse would have no effect whatsoever on the depolarization throughout the entire cell. In fact, for shunting inhibition, $G(x, x_I; t - t_I)$ is multiplied by

$$f_1 = a_E a_I V_E G(x_I, x_E; t_I),$$

whereas when the inhibition is of the nonshunting type $(V_I \neq 0)$, $G(x, x_I; t - t_I)$ is multiplied by

$$f_2 = a_I |V_I| \left[1 + |a_E V_E / V_I| G(x_I, x_E; t_I)\right].$$

Inserting numerical values for the parameters indicates that f_1 and f_2 can have similar magnitudes.

Koch et al. have solved numerically the following equation for the somatic potential in the case of shunting inhibition:

$$V_s(t) = \int_0^t \left[K_{es}(t - t')g_e(t')\{V_E - V(t')\} \right.$$

$$\left. - K_{is}(t - t')g_i(t')V(t')\right] dt',$$

where K_{es} and K_{is} are the Green's functions – that is, they give the somatic response for delta-function inputs at the excitatory and inhibitory input locations on a dendritic tree – and g_e and g_i are the conductance changes at those locations. To quantify the effectiveness of the shunting inhibition, Koch et al. define an F factor:

$$F = \frac{\text{maximum depolarization for excitation alone}}{\text{maximum depolarization for excitation followed by inhibition}}.$$

For the cable, the quantity F can be found from formula (7.99).

The calculations of Koch et al. were performed with data on ganglion cells of the cat retina with a view to understanding their direction sensitivity properties. The shunting-type inhibition was postulated as an underlying mechanism, with the following desirable characteristics:

(1) V_I must be close to zero and V_E must be large;
(2) the inhibitory synapse should be closer to the soma and on the same dendrite as the excitatory synapse (on path inhibition);
(3) the peak inhibitory conductance should be large; and
(4) excitation and inhibition should have considerable overlap.

Some further results are presented in Tuckwell (1985, 1986a) and Koch, Poggio, and Torre (1986).

7.5.3 Excitation (inhibition) followed by excitation (inhibition)

Suppose that an excitatory input of strength a_E arrives at x_E at $t = 0$, and there follows an excitatory input of strength a'_E at x'_E at

$t = t'$. That is, V satisfies

$$V_t = -V + V_{xx} + (V_E - V)\big[a_E \delta(x - x_E)\,\delta(t)$$
$$+ a'_E \delta(x - x'_E)\,\delta(t - t')\big],$$

with given boundary conditions and zero depolarization everywhere before the first input arrives. Then, using the same techniques as above, the response is found to be

$$V(x, t) = a_E V_E G(x, x_E; t) + a'_E V_E G(x, x'_E; t - t')$$
$$\times \big[1 - a_E G(x'_E, x_E; t')\big], \qquad t > t'.$$

Since $G(x'_E, x_E; t')$ is positive, it can be seen that the response to the second excitatory input is multiplied by a factor

$$r = 1 - a_E G(x'_E, x_E; t').$$

Some surprising results follow. If $r > 0$, so that

$$a_E < 1/G(x'_E, x_E; t'),$$

then the second EPSP is reduced in size but it is still excitatory. If, however, $r < 0$, or

$$a_E > 1/G(x'_E, x_E; t'),$$

then the second excitatory input will actually lead to a decrease in the level of depolarization (i.e., the EPSP will appear as an inhibitory synaptic potential). Furthermore, if $r = 0$, which occurs at the critical value

$$a_E = 1/G(x'_E, x_E; t'),$$

the second excitatory input has no effect whatsoever on the depolarization throughout the entire cell. It is as if the second input is not seen by the cell at all. The reason for this is that at the critical value of a_E, the value of the depolarization at the location and time of arrival of the second input, is exactly equal to the value of the excitatory reversal potential so that the net input current is zero. MacGregor (1968) alluded to these phenomena, with the remark that "a depolarizing synapse could be serving a primarily inhibitory function."

The same remarks apply to a pair of inhibitory inputs – in particular, there will be a critical value of a_I at which a second inhibitory input will not be seen at all.

7.5.4 Synaptic input with steady current at the origin
Let an excitatory impulsive conductance change, representing synaptic input, occur at the point x_0 at time $t = 0$. Then the voltage

satisfies

$$V_t = V_{xx} - V + a_E[V_E - V(x,t)]\,\delta(x - x_0)\,\delta(t), \qquad 0 < x < L.$$
$$\tag{7.101}$$

Suppose that prior to the arrival of the synaptic input, a constant current I_0 was injected at the origin (soma) and a steady-state distribution of potential, $U(x)$, $0 \le x \le L$, had been attained throughout the cell. To be specific, suppose a lumped-soma boundary condition is appropriate at $x = 0$ and a sealed-end condition applies at $x = L$. Thus

$$V_x(L,t) = 0, \tag{7.102}$$

$$V(0,t) + V_t(0,t) - \gamma V_x(0,t) = I_0 R_s, \tag{7.103}$$

and the initial condition is

$$V(x,0) = U(x). \tag{7.104}$$

The response is the same as if $V(x,t)$ satisfies the equation

$$V_t = V_{xx} - V + a_E[V_E - U(x_0)]\,\delta(x - x_0)\,\delta(t), \tag{7.105}$$

with the same boundary data. As seen in Section 6.5.2, the potential is given by

$$V(x,t) = U(x) + W(x,t), \tag{7.106}$$

where W satisfies

$$W_t = W_{xx} - W + a_E[V_E - U(x_0)]\,\delta(x - x_0)\,\delta(t), \tag{7.107}$$

$$W(0,t) + W_t(0,t) - \gamma W_x(0,t) = 0, \tag{7.108}$$

$$W_x(L,t) = 0. \tag{7.109}$$

That is, W is a constant multiple of the Green's function $G(x, x_0; t)$ for the cable equation, which was found in Section 6.3. Hence

$$V(x,t) = U(x) + a_E[V_E - U(x_0)]G(x, x_0; t). \tag{7.110}$$

If other boundary conditions are applied at $x = 0$ and $x = L$, then the Green's function to be used in this formula is the one appropriate to those boundary conditions.

Equation (7.110) gives, for the potential at $x = 0$ (soma, assumed site of recording electrode),

$$V_s(t) = U(0) + a_E[V_E - U(x_0)]G(0, x_0; t). \tag{7.111}$$

If $U(x_0) = V_E$ there will be no observable response at the soma to the synaptic input. Let the value of the potential at the origin, at which the response to an excitatory input becomes hyperpolarizing, be U_R. To find the true reversal potential, formula (6.17) for the steady-state potential under the application of the constant current I_0 at the origin must be employed. This gives, in the case of a sealed end at $x = L$,

$$V_E = U_R \cosh(L - x_0)/\cosh L, \qquad (7.112)$$

for the actual reversal potential. Thus a knowledge of both the location of the input and the electrotonic length of the nerve cylinder is needed to accurately obtain the reversal potential if the latter is judged from observations of the somatic response.

7.6 Response to sustained synaptic input: postsynaptic potentials

In the previous section the conductance increases during synaptic action were regarded as brief enough that they could be approximated as impulsive. That approach expedites the mathematical solution of the cable equation.

Transmitter action is, however, sustained as exemplified by certain synapses on spinal motoneurons (Burke and Rudomin 1977). The briefest Ia EPSPs are believed to be produced by conductance changes of duration 0.5 ms (about 0.1 time constants). EPSPs due to synapses from cells of the corticospinal tract are caused by somewhat longer conductance changes, and those due to activation of Ia inhibitory synapses last approximately 2 ms (about 0.4 time constants).

A more accurate representation of the changes in potential in response to, say, excitatory synaptic activation is obtained by solving

$$V_t = V_{xx} - V + a_E \delta(x - x_E)[H(t) - H(t - t_1)][V_E - V],$$

$$a < x < b. \quad (7.113)$$

Here the conductance increase lasts from $t = 0$ to $t = t_1$, a_E is a constant quantifying the magnitude of the change in conductance, and x_E is the location of the synapse. Assume that boundary conditions of a suitable type are given in $x = a$ and $x = b$ and that the cell is initially at rest. Then

$$V(x, 0) = 0, \qquad a \le x \le b. \qquad (7.114)$$

Let $G(x, y; t)$ be the Green's function for the cable equation $V_t = V_{xx} - V$, with the given boundary conditions. Then the potential is given by

$$V(x, t) = a_E \int_0^t G(x, x_E; t - s)[V_E - V(x_E, s)] \, ds, \qquad 0 \le t \le t_1.$$

$$(7.115)$$

The voltage distribution at $t = t_1$ is $V(x, t_1)$ and thereafter, since there are no sources of current, the potential is given by

$$V(x, t) = \int_a^b V(y, t_1) G(x, y; t - t_1) \, dy, \qquad t > t_1. \quad (7.116)$$

Equation (7.115) is an integral equation for V and may be solved in some instances using standard techniques. However, when the cable is infinite, an exact expression for $V(x, t)$ may be found using Laplace transforms.

7.6.1 Infinite cable: Laplace transform method

For the solution of Equation (7.113) on $-\infty < x < \infty$, we define the Laplace transform

$$V_L(x; p) = \int_0^\infty e^{-pt} V(x, t) \, dt, \qquad (7.117)$$

using the transform variable p since s is already taken. Application of the Laplace transformation operator to (7.115) and the use of the formula for the transform of a convolution integral gives, on rearranging,

$$V_L(x; p) = a_E G_L(x, x_E; p) [V_E/p - V_L(x_E; p)]. \quad (7.118)$$

Thus the Laplace transform of the solution at x involves the Laplace transform at x_E. The latter transform can be found explicitly as follows. Putting $x = x_E$ in (7.118) yields

$$V_L(x_E; p) = \frac{a_E V_E G_L(x_E, x_E; p)}{p[1 + a_E G_L(x_E, x_E; p)]}. \quad (7.119)$$

For the infinite cable

$$G_L(x_E, x_E; p) = \frac{1}{2(1 + p)^{1/2}}, \quad (7.120)$$

so

$$V_L(x_E; p) = \frac{a_E V_E}{p[a_E + 2(1 + p)^{1/2}]}. \quad (7.121)$$

Since

$$G_L(x, x_E; p) = \frac{\exp\{-|x - x_E| (1 + p)^{1/2}\}}{2(1 + p)^{1/2}}, \quad (7.122)$$

the required transform is

$$V_L(x; p) = \frac{a_E V_E}{2p(1 + p)^{1/2}} \exp\{-|x - x_E| (1 + p)^{1/2}\}$$

$$\times \left[1 - \frac{a_E}{a_E + 2(1 + p)^{1/2}}\right]. \quad (7.123)$$

To invert this transform we note that

$$\mathcal{L}\{\exp(t)V(x,t)\} = V_L(x; p-1)$$

$$= a_E V_E \exp\{-|x - x_E|\sqrt{p}\}$$

$$\times \left[\frac{1}{2(2 + a_E)(\sqrt{p} - 1)} + \frac{1}{2(2 - a_E)(\sqrt{p} + 1)} \right.$$

$$\left. + \frac{4}{(a_E^2 - 4)(a_E + 2\sqrt{p})} \right]. \qquad (7.124)$$

Making use of the inversion formula (Abramowitz and Stegun 1965, page 1026),

$$\mathcal{L}^{-1}\left\{ \frac{e^{-k\sqrt{p}}}{a + \sqrt{p}} \right\} = \frac{e^{-k^2/4t}}{\sqrt{\pi t}} - a e^{ak} e^{a^2 t} \operatorname{erfc}\left(a\sqrt{t} + \frac{k}{2\sqrt{t}} \right),$$

$$(7.125)$$

gives

$$V(x,t) = \left(\frac{a_E V_E}{2} \right) \left[\frac{\exp\{-|x - x_E|\}}{a_E + 2} \operatorname{erfc}\left(\frac{|x - x_E|}{2\sqrt{t}} - \sqrt{t} \right) \right.$$

$$+ \frac{\exp\{|x - x_E|\}}{a_E - 2} \operatorname{erfc}\left(\frac{|x - x_E|}{2\sqrt{t}} + \sqrt{t} \right)$$

$$+ \frac{2a_E}{4 - a_E^2} \exp\left\{ \frac{a_E |x - x_E|}{2} + \left(\frac{a_E^2}{4} - 1 \right) t \right\}$$

$$\left. \times \operatorname{erfc}\left(\frac{|x - x_E|}{2\sqrt{t}} + \frac{a_E \sqrt{t}}{2} \right) \right]. \qquad (7.126)$$

A similar result is given in Jack, Noble, and Tsien (1985). In the special case $a_E = 2$, the following result holds:

$$V(x,t) = \frac{V_E}{4} \left[\exp(-|x - x_E|) \operatorname{erfc}\left(\frac{|x - x_E|}{2\sqrt{t}} - \sqrt{t} \right) \right.$$

$$+ 4 \exp|x - x_E| \left\{ \sqrt{\frac{t}{\pi}} \exp\left[-\left(\frac{|x - x_E|}{2\sqrt{t}} + \sqrt{t} \right)^2 \right] \right.$$

$$\left. \left. - \left(\frac{1}{4} + \frac{|x - x_E|}{2} + t \right) \operatorname{erfc}\left(\frac{|x - x_E|}{2\sqrt{t}} + \sqrt{t} \right) \right\} \right]. \qquad (7.127)$$

The response at $x = x_E$ is

$$V(x_E, t) = \frac{a_E^2 V_E}{a_E^2 - 4}\left[1 - \frac{2}{a_E}\,\text{erf}(\sqrt{t}) - \exp\left\{\left(\frac{a_E^2}{4} - 1\right)t\right\}\text{erfc}\left(\frac{a_E\sqrt{t}}{2}\right)\right],$$

(7.128)

a formula given in Fatt and Katz (1951) and Burke (1957). When $a_E = 2$, this becomes

$$V(x_E, t) = \frac{V_E}{4}\left[\text{erfc}(-\sqrt{t}) + 4\left\{\sqrt{\frac{t}{\pi}}\,e^{-t} - \left(\frac{1}{4} + t\right)\text{erfc}(\sqrt{t})\right\}\right].$$

(7.128A)

The decay of potential

When the synaptically induced conductance increase is over, the potential decays according to the homogeneous cable equation with 'initial' value $V(x, t_1)$. From (7.115) this is given by

$$V(x, t_1) = a_E \int_0^{t_1} G(x, x_E; t_1 - s)[V_E - V(x_E, s)]\,ds.$$

(7.129)

Substituting in (7.116) gives

$$V(x, t) = a_E \int_{-\infty}^{\infty} \int_0^{t_1} G(y, x_E; t_1 - s)[V_E - V(x_E, s)]$$
$$\times G(x, y; t - t_1)\,ds\,dy, \qquad t > t_1.$$

(7.130)

Using formula (7.79) this simplifies to

$$V(x, t) = a_E \int_0^{t_1} G(x, x_E; t - s)[V_E - V(x_E, s)]\,ds. \quad (7.131)$$

The first part of this integral can be evaluated using (5.48) to give

$$V(x, t) = \frac{a_E V_E}{4}\left[e^{-|x - x_E|}\left\{\text{erfc}\left(\frac{|x - x_E| - 2t}{2\sqrt{t}}\right)\right.\right.$$
$$\left. - \text{erfc}\left(\frac{|x - x_E| - 2(t - t_1)}{2\sqrt{t - t_1}}\right)\right\}$$
$$- e^{|x - x_E|}\left\{\text{erfc}\left(\frac{|x - x_E| + 2t}{2\sqrt{t}}\right)\right.$$
$$\left.\left. - \text{erfc}\left(\frac{|x - x_E| + 2(t - t_1)}{2\sqrt{t - t_1}}\right)\right\}\right]$$
$$- a_E \int_0^{t_1} G(x, x_E; t - s)V(x_E, s)\,ds, \qquad t > t_1. \quad (7.132)$$

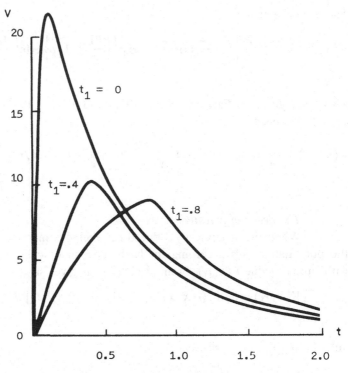

Figure 7.9A. EPSPs recorded at $x = 0$ for an impulsive conductance change ($t_1 = 0$) and for various durations of the synaptically induced conductance change.

The last integral here can be reduced to integrals of bivariate normal densities but is probably best evaluated numerically as a one-dimensional integral since $V(x_E, s)$ is readily obtainable from (7.128) for $0 \leq s \leq t_1$.

Computed EPSPs for the infinite cable are shown in Figure 7.9A. The largest EPSP corresponds to the case of an impulsive conductance change [formula (7.76)]. The remaining EPSPs are calculated from (7.126) and (7.132), where the normalization is $a_E t_1 = 1$ and t_1 is the duration of the conductance increase. In all cases the potential is given at $x = 0$ with the synaptic input located at $x_E = 0.5$, the value of V_E being 50 mV. The peak voltage is plotted against t_1 in Figure 7.9B. The results indicate the limitations of approximating a sustained synaptic input with an impulsive one.

7.6.2 Finite cables

When the nerve cylinder has a finite length, the response to a sustained conductance increase is more difficult to obtain by the

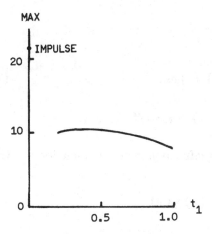

Figure 7.9B. Peak voltage during an EPSP plotted against the duration of the conductance change producing it. For a further explanation, see the text.

Laplace transform method. An alternative method of solution is nevertheless available. Equation (7.115), at the location of the synapse, is

$$V(x_E, t) = a_E V_E \int_0^t G(x_E, x_E; t-s) \, ds$$

$$- a_E \int_0^t G(x_E, x_E; t-s) V(x_E, s) \, ds. \qquad (7.133)$$

This equation is called a *Volterra equation* and may be solved, for example, by the method of *successive substitution* (Lovitt 1950). To standardize the notation, we put

$$u(t) = V(x_E, t), \qquad (7.134)$$

$$f(t) = a_E V_E \int_0^t G(x_E, x_E; t-s) \, ds, \qquad (7.135)$$

$$K(t, s) = -G(x_E, x_E; t-s), \qquad (7.136)$$

so (7.133) becomes

$$u(t) = f(t) + a_E \int_0^t K(t, s) u(s) \, ds. \qquad (7.137)$$

Successive substitution of $u(t)$ in the right-hand side yields

$$u(t) = \sum_{n=0}^{\infty} a_E^n T_n(t), \qquad (7.138)$$

where

$$T_0(t) = f(t), \tag{7.139}$$

$$T_1(t) = \int_0^t K(t, s) f(s) \, ds, \tag{7.140}$$

$$T_2(t) = \int_0^t \int_0^s K(t, s) K(s, t_1) f(t_1) \, dt_1 \, ds. \tag{7.141}$$

If we use the general eigenfunction representation for the Green's function

$$G(x, y; t) = \sum_n \phi_n(x) \phi_n(y) e^{-\mu_n t}, \tag{7.142}$$

then the leading term is

$$T_0(t) = a_E V_E \sum_n \phi_n^2(x_E)[1 - e^{-\mu_n t}], \tag{7.143}$$

and the next term is

$$T_1(t) = a_E V_E \left[\sum_n \frac{\phi_n^4(x_E) e^{-\mu_n t}}{\mu_n} \left(\frac{e^{\mu_n t} - 1}{\mu_n} - t \right) \right.$$

$$+ \sum_m \sum_{\substack{n \\ m \neq n}} \frac{\phi_m^2(x_E) \phi_n^2(x_E)}{\mu_n}$$

$$\left. \times \left(\frac{1 - e^{-\mu_m t}}{\mu_m} - \frac{e^{-\mu_m t} \{ e^{(\mu_m - \mu_n) t} - 1 \}}{\mu_m - \mu_n} \right) \right]. \tag{7.144}$$

Higher-order terms in the expansion may also be evaluated.

To obtain the potential along the entire cylinder, the value of $V(x_E, t)$ is inserted in (7.115). Define

$$U(x, t) = a_E V_E \int_0^t G(x, x_E; t - s) \, ds, \tag{7.145}$$

$$V_n(x, t) = a_E \int_0^t G(x, x_E; t - s) T_n(s) \, ds. \tag{7.146}$$

Then

$$V(x, t) = U(x, t) - \sum_{n=0}^{\infty} V_n(x, t). \tag{7.147}$$

The first few terms are

$$U(x,t) = a_E V_E \sum_n \phi_n(x)\phi_n(x_E)(1 - e^{-\mu_n t}), \qquad (7.148)$$

$$V_o(x,t) = a_E^2 V_E \left[\sum_n \frac{\phi_n(x)\phi_n^3(x_E)e^{-\mu_n t}}{\mu_n} \left(\frac{e^{\mu_n t} - 1}{\mu_n} - t \right) \right.$$

$$+ \sum_n \sum_{\substack{m \\ n \neq m}} \phi_n(x)\phi_n(x_E)e^{-\mu_n t} \frac{\phi_m^2(x_E)}{\mu_m}$$

$$\left. \times \left(\frac{e^{\mu_n t} - 1}{\mu_n} - \frac{\{e^{(\mu_n - \mu_m)t} - 1\}}{\mu_n - \mu_m} \right) \right].$$

$$(7.149)$$

Higher-order terms may be similarly evaluated.

7.7   Neurons with dendritic trees

When the cable equation with reversal potentials is employed in order to find the depolarization over neurons with dendritic trees, much less progress can be, or has been, made in finding exact solutions. For example, as the reader may convince himself, the mapping procedure of Chapter 5 is no longer applicable. Koch and Poggio (1985), however, have made considerable progress by Laplace transforming the problems and adopting the approach of Butz and Cowan (see Chapter 5). Since we have seen that the inclusion of synaptic reversal potentials leads to interaction phenomena, which do not occur with the fully linear cable equation, this is an area in which much further work is needed. We will make a brief examination of some of the problems that arise.

The equivalent cylinder

The equivalent-cylinder concept extends to neurons with dendritic trees when the potential on each dendrite satisfies an equation of the kind

$$V_t = -V + V_{xx} + (V_E - V)g_E(x,t) + (V_I - V)g_I(x,t),$$

providing there is sufficient symmetry such that the potential and synaptic conductance terms are the same at all points that are the same electrotonic distance from the origin or soma. In addition, the three-halves power law for diameters must be obeyed at each branch point. The proof of this can be styled on that given for the "usual" cable equation, except that Laplace transforms are not useful in converting the partial differential equations to ordinary ones in most cases.

A neuron initially at rest

Suppose the potential throughout a whole neuron with one or more dendritic trees is initially at resting level and that at $t = 0$ an arbitrary number of impulsive (excitatory or inhibitory or both) conductance changes occur at various locations, representing simultaneous multi-synaptic activation. Because the initial depolarization is zero, the V's in the terms $(V_E - V)g_E$ and $(V - V_I)g_I$ have no effect and the response of the cell is the same as it would be if these terms were just $V_E g_E$ and $V_I g_I$. This was exemplified in Section 7.5.1, where the response before the second excitatory input arrived was simply $a_E V_E G(x, x_E; t)$. Thus, when the cell is initially at rest and *impulsive* conductance changes occur simultaneously, the techniques of Chapter 6, Section 7, may be employed. However, if there are inputs at different times, this is no longer true and the boundary-value problems for the partial differential equations including the reversal potentials must be solved. This produces very complicated problems as even the following simple example illustrates.

A dendritic tree with one branch point

Consider the dendritic tree sketched in Figure 7.10. Let the space coordinates be as shown with $x \in (0, L_1)$, $y \in (0, L_2)$, and $z \in (0, L_3)$. The initial distribution of depolarization is not assumed to be zero; that is,

$$U(x,0) = u(x),$$
$$V(y,0) = v(y),$$
$$W(z,0) = w(z).$$

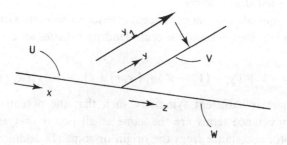

Figure 7.10. Dendritic tree for text example.

Suppose an impulsive excitatory conductance change occurs at $t = 0$ at a point a distance y_1 from the branch point on the upper right-hand dendrite. The governing differential equations are then

$$U_t = -U + U_{xx}, \qquad\qquad x \in (0, L_1), \; t > 0,$$

$$V_t = -V + V_{yy} + a_E(V_E - V)\delta(t)\delta(y - y_1), \qquad y \in (0, L_2), \; t > 0,$$

$$W_t = -W + W_{zz}, \qquad\qquad z \in (0, L_3), \; t > 0.$$

Assume that dendritic terminals are sealed so that there are the boundary conditions

$$U_x(0, t) = V_y(L_2, t) = W_z(L_3, t) = 0.$$

The additional constraints are those of continuity of potential

$$U(L_1, t) = V(0, t) = W(0, t),$$

and conservation of axial current

$$U_x(L_1, t)/\bar{r}_1 = V_y(0, t)/\bar{r}_2 + W_z(0, t)/\bar{r}_3,$$

where $\bar{r}$ is the axial resistance per characteristic length.

To find the potential throughout the tree one may use Laplace transforms. Denoting these by bars and using s as the transform variable, we have the following ordinary differential equations:

$$-\bar{U}'' + (s+1)\bar{U} = u(x), \qquad\qquad x \in (0, L_1),$$

$$-\bar{V}'' + (s+1)\bar{V} = v(y) + a_E\delta(y - y_1)(V_E - v(y)),$$

$$\qquad\qquad y \in (0, L_2),$$

$$-\bar{W}'' + (s+1)\bar{W} = w(z), \qquad\qquad z \in (0, L_3),$$

$$\tag{7.150}$$

with the constraints

$$\bar{U}'(0; s) = \bar{V}'(L_2; s) = \bar{W}'(L_3; s) = 0,$$

$$\bar{U}(L_1; s) = \bar{V}(0; s) = \bar{W}(0; s), \tag{7.151}$$

$$\bar{U}'(L_1; s)/\bar{r}_1 = \bar{V}'(0; s)/\bar{r}_2 + \bar{W}'(0; s)/\bar{r}_3.$$

These equations may be solved using the techiques of Chapter 4.

8

Theory of the action potential

8.1   Introduction

In Chapter 1 we saw that sufficient depolarization of a motoneuron, due either to current injection or excitatory synaptic input, could cause the occurrence of an *action potential*. During the action potential the polarity of the cell membrane was briefly reversed, the inside of the cell becoming temporarily at a positive electrical potential relative to the outside. Furthermore, this depolarizing pulse propagated along the axon of the cell to eventually reach the target muscle cells that were excited at the neuromuscular synapses. The reader should also review the first experimental recording of the action potential, that for the squid giant axon shown in Figure 2.1.

In this chapter a theory of the action potential will be presented. This theory relies almost solely on the works of the great trio of physiologists, Hodgkin, Huxley, and Katz. Not only did they perform the many key experiments but they also developed cogent mathematical models in the form of systems of differential equations – the Hodgkin–Huxley equations. These equations have given considerable impetus to the mathematical study of nonlinear partial differential equations. The equations cannot be solved in the sense that solutions of the linear (Chapters 5 and 6) and bilinear (Chapter 7) cable equations could be found. Hence solutions must be obtained by numerical methods and these will be presented in detail. Although the lack of availability of analytic solutions is cumbersome, it will be seen that what is lost in mathematical tractability is more than compensated by the gain in power to describe physiological reality. In particular, the action potential and the *threshold conditions* for its instigation are a natural property of the equations.

8.2 Ionic currents and equilibrium potentials

In Chapter 2 we noted that there is usually a much greater concentration of potassium ions in the intracellular compartment of nerve cells than in the extracellular fluid and that for sodium and chloride ions the situation was reversed. The Goldman–Hodgkin–Katz formula (see Section 2.10.1) predicts that the membrane potential is given by

$$V_m = \frac{RT}{F} \ln \left[\frac{P_K[K]_0 + P_{Na}[Na]_0 + P_{Cl}[Cl]_i}{P_K[K]_i + P_{Na}[Na]_i + P_{Cl}[Cl]_0} \right]. \tag{8.1}$$

In the resting state the ratios of the values $P_K : P_{Na} : P_{Cl}$ are approximately $1 : 0.04 : 0.4$ (for the squid axon) and when these values and the appropriate internal and external ion concentrations are inserted in (8.1), a resting potential of about -60 mV is obtained in agreement with experiment.

It was suggested (Hodgkin and Katz 1949) that during the action potential the permeability of the membrane to Na^+ increases transiently (activation followed by inactivation) to become temporarily greater than the permeabilities to both K^+ and Cl^-. This "sodium hypothesis" was attractive because, in the extreme case of a membrane permeable only to Na^+, the membrane potential predicted by (8.1) would be that of the sodium Nernst potential V_{Na}. To test this idea, Hodgkin and Katz reduced the external sodium concentration from its usual value and found a reduction occurred in the magnitude of the action potential. The resting membrane potential was hardly affected. Similarly, increasing the external sodium concentration led to an increase in the magnitude of the action potential. It was found that conduction of the nerve impulse was impossible unless Na^+, or a suitable substitute ion (Li^+), was present in the external fluid. Also, chloride was not essential for action potentials since its replacement by other anions did not reduce the ability of the cell to conduct impulses.

Evidence had also accumulated that K^+ was extruded from cells during nervous activity. An additional hypothesis was subsequently made (Hodgkin 1951) that not only was there a transient increase in sodium permeability but that there was a subsequent increase in potassium permeability. This would lead to a faster repolarization of the membrane. *Refractoriness* of the nerve following an impulse could be explained by the increase in P_K and the suppression of P_{Na} during the final stages of the spike.

If these ideas are correct, then during the action potential, the potential should swing first from resting level toward V_{Na}, then toward V_K, and finally return to resting level. This is indeed the observed sequence of events (cf. Figure 1.26). These are the principal ideas behind the theory of the action potential. A detailed experimental investigation and theoretical analysis were subsequently performed by Hodgkin, Huxley, and Katz (Hodgkin, Huxley, and Katz 1952; Hodgkin and Huxley 1952a–d).

A major achievement in these experiments was the division of the current through the membrane into the components of capacitative and ionic currents. For a small patch of membrane the total current is assumed to be

$$I = C\,dV/dt + I_i, \tag{8.2}$$

where $C\,dV/dt$ is the capacitative current and I_i is the ionic current. The experimental technique of voltage-clamping (Cole 1949; Marmont 1949) was the key in separating the components. In this technique, the voltage V is held constant across a patch of membrane by means of a feedback current. Since $dV/dt = 0$, all capacitative current is eliminated and the ionic current can be measured.

The voltage-clamp technique was applied to the squid axon by Hodgkin, Huxley, and Katz (1952). In a typical experiment, if the membrane was held at a depolarized level, the ionic current was at first inward (positive) and smoothly changed into a persistent outward current. The next problem was to separate the ionic current into *its* components according to the ionic species that contribute to it.

Hodgkin and Huxley (1952a) used choline to replace the Na^+ in the external fluid. The inward current, assumed carried by Na^+, was abolished, leaving the outward component only, assumed carried by K^+. The total ionic current was expressed as

$$I_{ion} = I_K + I_{Na} + I_l, \tag{8.3}$$

where I_K and I_{Na} are the potassium and sodium currents, and I_l is a "leakage" current carried by other ions (e.g., chloride).

It was further posited that the "driving force" for an ion species was proportional to the difference between the membrane potential and the Nernst potential of the ion species. Thus, for Na^+, the permeability of the membrane was measured by* $I_{Na}/(V - V_{Na})$ and this quantity was *defined* as the sodium conductance g_{Na}. Similarly,

---

*Henceforth voltages will be relative to the resting potential.

for the other ionic components, so that we have

$$g_{Na} = I_{Na}/(V - V_{Na}),$$ (8.4)

$$g_K = I_K/(V - V_K),$$ (8.5)

$$g_l = I_l/(V - V_l).$$ (8.6)

With the voltage-clamp technique the time courses of g_{Na} and g_K at various voltages could be determined. In a typical experiment, holding the membrane at a depolarized level would cause a transient increase in g_{Na} and a slowly rising increase in g_K to some steady-state value. Furthermore, if the membrane was held at a depolarized level for some time and a further depolarization applied, the inward current was less than it would have been if just the final depolarization had been applied. Thus there were two factors governing the sodium conductance – one that increased it (the *activation* process) and one that decreased it (the sodium *inactivation* process). On the other hand, depolarizing the membrane only led to an increase in the potassium conductance.

It is of interest to note that the original method of determination of the ionic components was to substitute choline for Na^+ in the extracellular fluid. The choline ions are too large to pass through the ion channels so the inward current is eliminated. The separation of the ionic current into its components may now be achieved by the use of pharmacological agents that alter the membrane permeability to various ions. *Tetrodotoxin* (TTX), obtained from the Japanese puffer fish (and long used in Chinese herb recipes), abolishes the inward Na^+ current without affecting the resting potential (Nakamura, Nakajima, and Grundfest 1965). It has been demonstrated that this blocking of the Na^+ current is by the binding of TTX molecules to receptors in the membrane. *Tetraethylammonium chloride* (TEA) applied to the interior of cells efficiently blocks the outward K^+ current as demonstrated, for example, by Hagiwara and Saito (1959). An important consequence of these pharmacological effects is the demonstration that drugs block K^+ and Na^+ currents separately, which supports the idea that the K^+ and Na^+ currents are *independent* (Hodgkin and Huxley 1952a) and occur through specialized *channels* in the membrane (cf. Section 2.2).

The electrical circuit employed by Hodgkin and Huxley (1952d) to represent a patch of membrane is shown in Figure 8.1. The membrane capacitance C is in parallel with resistances R_{Na}, R_K, and R_l, which are the reciprocals of the corresponding conductances defined in (8.4)–(8.6). The driving potentials across these resistances are $V - V_{Na}$,

outside

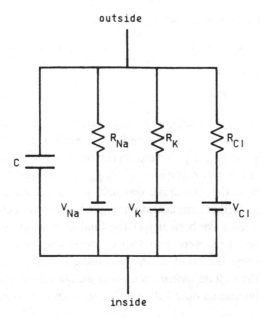

inside

Figure 8.1. The electrical circuit used by Hodgkin and Huxley to represent a patch of nerve membrane.

$V - V_K$, and $V - V_l$, respectively. The total current is then

$$I = C\,dV/dt + g_{Na}(V - V_{Na}) + g_K(V - V_K) + g_l(V - V_l).$$
(8.7)

Note that at rest, with $dV/dt = 0$ and $I = 0$, the value of the potential predicted by (8.7) is

$$V_R = (V_{Na}g_{Na} + V_Kg_K + V_lg_l)/(g_{Na} + g_K + g_l),$$
(8.8)

which differs from that given by (8.1).

8.3 Quantitative description of the potassium and sodium conductances

Figures 8.2A and 8.2B show the time courses of the potassium and sodium conductances of the squid axon obtained under a voltage clamp applied at $t = 0$. Each curve is marked with the depolarization of the clamp relative to resting level (numbers on the right in Figure 8.2A, on the left in Figure 8.2B).

8.3.1 The potassium conductance g_K

Under a depolarizing voltage clamp, g_K increases to smoothly approach an asymptotic steady-state value. It will be noted from Figure 8.2A that greater depolarizations lead to larger asymptotic values and faster rates of approach to them.

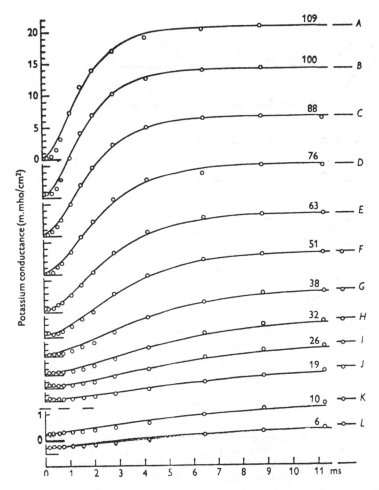

Figure 8.2A. Time courses of the potassium conductance of the squid axon when the membrane is clamped at various voltages. The numbers on the right of each curve are the depolarizations (in millivolts). Note the change in ordinate scale. [From Hodgkin and Huxley (1952d). Reproduced with the permission of The Physiological Society and the authors.]

Before proceeding, we pause to consider the first-order linear differential equation

$$dy/dt = \alpha(1 - y) - \beta y, \qquad t > 0, \; y(0) = y_0. \tag{8.9}$$

By the method of solution given in Section 2.4, we obtain

$$y(t) = \frac{\alpha}{\alpha + \beta}\left[1 + \left\{\left(\frac{\alpha + \beta}{\alpha}\right)y_0 - 1\right\}e^{-(\alpha + \beta)t}\right]. \tag{8.10}$$

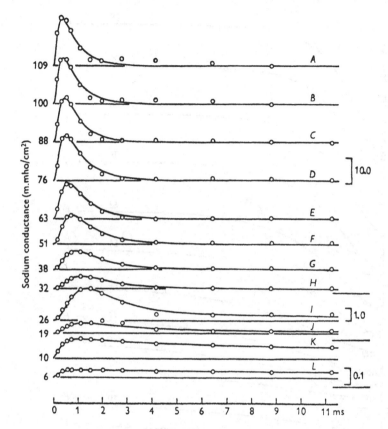

Figure 8.2B. Time courses of the sodium conductance at various clamped voltages. Note the change in ordinate scales. [From Hodgkin and Huxley (1952d). Reproduced with the permission of The Physiological Society and the authors.]

As $t \to \infty$, $y(t) \to y_\infty$, given by

$$y_\infty = \alpha/(\alpha + \beta). \tag{8.11}$$

If we introduce the *time constant*

$$\tau = 1/(\alpha + \beta), \tag{8.12}$$

the solution (8.10) can be written as

$$y(t) = y_\infty + (y_0 - y_\infty)e^{-t/\tau}. \tag{8.13}$$

Solutions will appear, for appropriate α, β, and y_0, as in Figure 8.3.

Under voltage clamp, the behavior of g_K is of the same qualitative behavior as the solutions of (8.9). Both the asymptote and the time constant of the conductance depend on V, so the coefficients α and β

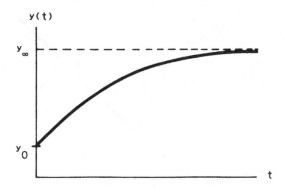

Figure 8.3. Graph of the solution of (8.9).

should be functions of V. It was found that a good fit to the g_K-versus-t curves in Figure 8.2A could be obtained by setting

$$g_K = \bar{g}_K n^4, \tag{8.14}$$

where $\bar{g}_K$ is a constant conductance and n satisfies a differential equation of the form of (8.9). The quantity n is referred to as the *potassium activation variable*, is dimensionless, takes values in $[0,1]$, and satisfies, at fixed V,

$$dn/dt = \alpha_n(1-n) - \beta_n n, \qquad t > 0, \; n(0) = n_0. \tag{8.15}$$

At fixed V the solution of this equation is

$$n(t) = n_\infty + (n_0 - n_\infty)e^{-t/\tau_n}, \tag{8.16}$$

where

$$n_\infty = \alpha_n/(\alpha_n + \beta_n), \tag{8.17}$$

$$\tau_n = 1/(\alpha_n + \beta_n). \tag{8.18}$$

The dependence of the constants α_n and β_n on V was found empirically from the g_K-versus-t curves to be satisfactorily approximated by

$$\alpha_n(V) = \frac{(10 - V)}{100[e^{(10-V)/10} - 1]}, \tag{8.19}$$

$$\beta_n(V) = \tfrac{1}{8}e^{-V/80}. \tag{8.20}$$

These functions appear as sketched in Figure 8.4. Note that α_n is considerably larger than β_n over most of the range of physiological voltages and, whereas α_n increases with increasing depolarization, β_n decreases.

Some insight into the nature of changes in g_K when V changes may be obtained by considering the following. If, with the voltage held at

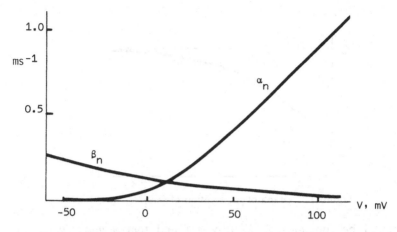

Figure 8.4. The dependence of the coefficients α_n and β_n on the depolarization V. The curves are redrawn from those of Hodgkin and Huxley (1952d).

V_0, the potassium conductance has reached a steady-state value at $t = t_0$, then the corresponding value of n must be

$$n_\infty(V_0) = \frac{\alpha_n(V_0)}{\alpha_n(V_0) + \beta_n(V_0)}. \tag{8.21}$$

If V is held at V_0 until $t = t_1$, n will remain at this value. An abrupt increase in V to the value $V_1 > V_0$ at t_1 will lead to an increase in n toward the new value

$$n_\infty(V_1) = \frac{\alpha_n(V_1)}{\alpha_n(V_1) + \beta_n(V_1)}, \tag{8.22}$$

and, whilst the clamp voltage is V_1, with $\tau_1 = 1/(\alpha_n(V_1) + \beta_n(V_1))$,

$$n(t) = n_\infty(V_1) + (n_\infty(V_0) - n_\infty(V_1))e^{-(t-t_1)/\tau_1}. \tag{8.23}$$

A subsequent downward step in V to V_2 at $t = t_2$ will give, for $t > t_2$,

$$n(t) = n_\infty(V_2) + (n(t_2) - n_\infty(V_2))e^{-(t-t_2)/\tau_2}, \tag{8.24}$$

where $\tau_2 = 1/(\alpha_n(V_2) + \beta_n(V_2))$. The trajectory of $n(t)$ is sketched in Figure 8.5.

8.3.2   The sodium conductance g_{Na}

Two variables are needed to describe the behavior of the sodium conductance. One is an *activation variable m*; the other is an *inactivation variable h*. In terms of these the sodium conductance is fitted by

$$g_{Na} = \bar{g}_{Na}m^3h, \tag{8.25}$$

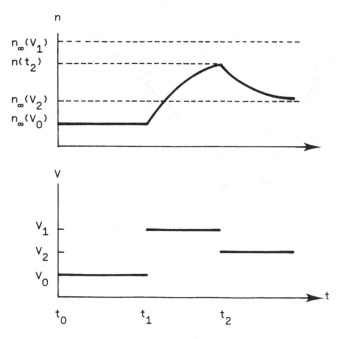

Figure 8.5. The behavior of the potassium activation variable $n(t)$ when the voltage is suddenly changed, first in the depolarizing and second in the hyperpolarizing direction.

where $\bar{g}_{Na}$ is a constant conductance, and m and h are dimensionless quantities taking values in $[0,1]$. Both m and h satisfy equations such as (8.9):

$$dm/dt = \alpha_m(1-m) - \beta_m m, \qquad (8.26)$$

$$dh/dt = \alpha_h(1-h) - \beta_h h, \qquad (8.27)$$

where the coefficients depend on voltage as follows:

$$\alpha_m(V) = \frac{25-V}{10[e^{(25-V)/10}-1]}, \qquad (8.28)$$

$$\beta_m(V) = 4e^{-V/18}, \qquad (8.29)$$

$$\alpha_h(V) = \tfrac{7}{100}e^{-V/20}, \qquad (8.30)$$

$$\beta_h(V) = \frac{1}{e^{(30-V)/10}+1}. \qquad (8.31)$$

These functional relationships are shown in Figures 8.6 and 8.7.

Solutions of (8.26) and (8.27) for initial values m_0 and h_0 will be given, at fixed V, by

$$m(t) = m_\infty + (m_0 - m_\infty)e^{-t/\tau_m}, \qquad (8.32)$$

$$h(t) = h_\infty + (h_0 - h_\infty)e^{-t/\tau_h}, \qquad (8.33)$$

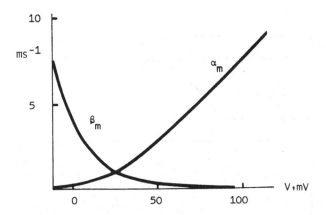

Figure 8.6. The dependence of α_m and β_m on depolarization.

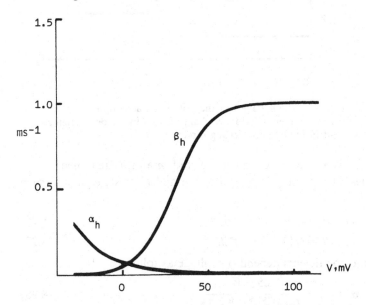

Figure 8.7. The dependence of α_h and β_h on depolarization.

where

$$m_\infty = \frac{\alpha_m}{\alpha_m + \beta_m}, \tag{8.34}$$

$$\tau_m = \frac{1}{\alpha_m + \beta_m}, \tag{8.35}$$

$$h_\infty = \frac{\alpha_h}{\alpha_h + \beta_h}, \tag{8.36}$$

$$\tau_h = \frac{1}{\alpha_h + \beta_h}. \tag{8.37}$$

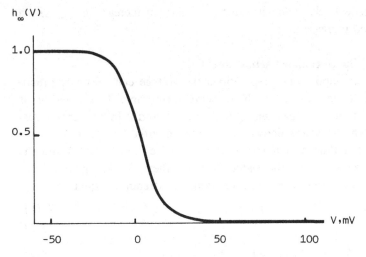

Figure 8.8. Asymptotic values of the sodium inactivation variable h as a function of depolarization.

If there were no sodium inactivation variable, the sodium conductance would behave in a similar fashion to the potassium conductance. The sodium conductance is brought down eventually by virtue of the fact that the asymptotic values of $h(t)$ are extremely small for most relevant values of V. This is shown in Figure 8.8.

Suppose now the membrane is clamped in the resting state. From the formulas for n_∞, m_∞, and h_∞, the values of n, m, and h are about 0.3, 0.05, and 0.6, respectively. Under a depolarizing clamp, n and m increase, whereas h decreases as sketched in Figure 8.9. For the case illustrated, the final values of n, m, and h are 0.72, 0.65, and

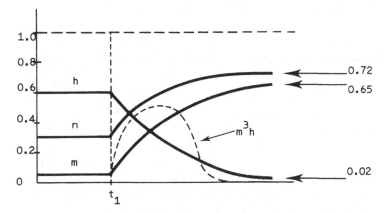

Figure 8.9. Temporal variation of n, m, and h when a depolarizing voltage clamp of 30 mV is applied at $t = t_1$. The dotted line gives the product m^3h, which is proportional to the sodium conductance.

0.02, respectively. The net effect is a transient increase in g_{Na} and a sustained increase in g_K.

8.4   Space-clamped action potential

In laboratory preparations the voltage can be made practically uniform over a patch of nerve membrane. This situation is referred to as a *space clamp* (cf. Chapter 7) and it follows that across the patch the space derivatives of the potential are zero. Diffusive effects can thus be ignored. If the applied current is $I_A(t)$ and the depolarization over the patch is $V(t)$, then the Hodgkin–Huxley equations are the system of four ordinary differential equations:

$$C\,dV/dt + \bar{g}_K n^4 (V - V_K) + \bar{g}_{Na} m^3 h (V - V_{Na}) + g_l (V - V_l) = I_A(t),$$
$$(8.38)$$

$$dn/dt = \alpha_n (1 - n) - \beta_n n, \qquad (8.39)$$
$$dm/dt = \alpha_m (1 - m) - \beta_m m, \qquad (8.40)$$
$$dh/dt = \alpha_h (1 - h) - \beta_h h. \qquad (8.41)$$

Consider the effect of delivering an impulsive current at $t = 0$ so that

$$I_A(t) = Q\,\delta(t), \qquad (8.42)$$

where Q is the charge delivered to the membrane and $\delta(t)$ is the delta function (see Section 2.10.1). If I_A is a delta function at $t = 0$, then, from (8.38), so is dV/dt. If the derivative is a delta function, then the integral of this, the function itself, must have a step discontinuity. Thus V has a jump at $t = 0$ to say V_0, which depends on Q. Also, the rate coefficients, the α's and β's, will also jump to their values at V_0. Then we may consider instead the initial-value problem

$$C\,dV/dt + \bar{g}_K n^4 (V - V_K) + \bar{g}_{na} m^3 h (V - V_{Na}) = 0, \qquad V(0) = V_0,$$
$$(8.38')$$

in conjunction with Equations (8.39)–(8.41). The initial values of n, m, and h are their resting steady-state values $[n(0) = \alpha_n(0)/(\alpha_n(0) + \beta_n(0))$, etc., where the zero argument for α_n and β_n refers to $V = 0]$. In these calculations the leakage current is usually ignored.

If V_0 is large enough, an action potential may develop. This is called a *space-clamped action potential* or *membrane action potential* by Hodgkin and Huxley. Some of their calculated results are shown in Figure 8.10 as well as some closely related experimental results. Hodgkin and Huxley used a method of integrating the system (8.38') and (8.39)–(8.41) given by Hartree in 1932. An efficient method of integration will be given later in this section.

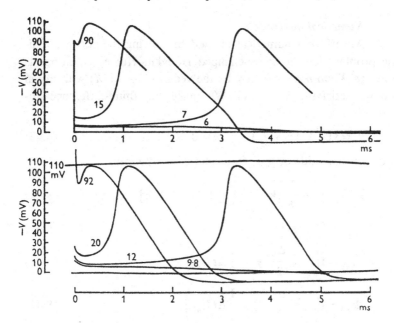

Figure 8.10. Depolarization as a function of time in the space-clamped Hodgkin–Huxley equations for various initial voltages. The upper set of results is calculated and the lower set is from experiment. [From Hodgkin and Huxley (1952d). Reproduced with the permission of The Physiological Society and the authors.]

In Figure 8.10 the calculated membrane depolarization returns slowly to zero and actually a hyperpolarization follows, which may be due to the fact that $V = 0$ is not an equilibrium point or due to accumulated numerical error. For a slightly larger initial depolarization (7 mV), $V(t)$ at first increases slowly and eventually climbs to about 100 mV. Thereafter $V(t)$ returns to zero and overshoots it to give a long-lasting *afterhyperpolarization*. This trajectory is identified as a space-clamped action potential. The theoretical *threshold initial depolarization* is therefore between 6 and 7 mV. Note that there is a discrepancy of a few millivolts (possibly greater than 3 mV) between the theoretical and experimental threshold voltages.

Examination of the trajectories $\{V(t),\ t \geq 0\}$, for $V_0 = 15$ mV and $V_0 = 90$ mV reveals that although dV/dt is initially negative, an action potential develops whose peak is almost identical to that for $V_0 = 7$ mV. Furthermore, the falling edge of the action potential is almost identical in the three cases $V_0 = 7$, 15, and 90 mV. This invariance leads to the use of the term "all or none" with reference to action potentials.

Numerical integration

An efficient numerical method for solving the above initial-value problem for the space-clamped Hodgkin–Huxley equations is the *Runge–Kutta method*. Let $\mathbf{y}$ be the vector (V, n, m, h) with initial value $\mathbf{y}_0$, satisfying $\dot{\mathbf{y}} = \mathbf{f}(t, \mathbf{y})$. We make the finite-difference approximation

$$\mathbf{y}_n \simeq \mathbf{y}(n\,\Delta t) \doteq \mathbf{y}(t_n), \qquad n = 0, 1, 2, \ldots, \tag{8.43}$$

and obtain the value of $\mathbf{y}_{n+1}$ from that of $\mathbf{y}_n$ by means of

$$\mathbf{y}_{n+1} = \mathbf{y}_n + \frac{\Delta t}{6}[\mathbf{k}_{n1} + 2\mathbf{k}_{n2} + 2\mathbf{k}_{n3} + \mathbf{k}_{n4}] + o(\Delta t^5), \tag{8.44}$$

$$\mathbf{k}_{n1} = \mathbf{f}(t_n, \mathbf{y}_n), \tag{8.45}$$

$$\mathbf{k}_{n2} = \mathbf{f}\left(t_n + \frac{\Delta t}{2}, \mathbf{y}_n + \frac{1}{2}\Delta t\,\mathbf{k}_{n1}\right), \tag{8.46}$$

$$\mathbf{k}_{n3} = \mathbf{f}\left(t_n + \frac{\Delta t}{2}, \mathbf{y}_n + \frac{1}{2}\Delta t\,\mathbf{k}_{n2}\right), \tag{8.47}$$

$$\mathbf{k}_{n4} = \mathbf{f}(t_n + \Delta t, \mathbf{y}_n + \Delta t\,\mathbf{k}_{n3}). \tag{8.48}$$

This is the vector form of the Runge–Kutta formula given in Boyce and DiPrima (1977). Note that reducing Δt by $1/2$ reduces the local error by a factor of $1/32$. As an exercise the reader may write a computer program to solve the space-clamped Hodgkin–Huxley equations using the Runge–Kutta formula. The method has been employed by Goldstein and Rall (1974) on a somewhat simplified system of equations to study the changes in action-potential form as it moves on an inhomogeneous axon. Most large computing facilities have a Runge–Kutta package, which has inbuilt refinements such as error control.

8.5 Propagating action potential: traveling-wave equations

In the cable model of Section 4.2, we replace the R–C circuits representing elemental lengths of membrane by Hodgkin–Huxley-type circuits as in Figure 8.1. We then obtain

$$c_m \frac{\partial V}{\partial t} = \frac{1}{r_i}\frac{\partial^2 V}{\partial x^2} + g_K^*(V_K - V) + g_{Na}^*(V_{Na} - V)$$

$$+ g_l^*(V_l - V) + I_A^*, \tag{8.49}$$

where V is the depolarization, c_m is the membrane capacitance per unit length, r_i is the axial resistance per unit length, g_K^*, g_{Na}^*, and g_l^* are the potassium, sodium, and leakage conductances per unit length, respectively, and I_A^* is the applied current density (per unit length). If the fiber radius is a, then

$$c_m = 2\pi a C_m, \tag{8.50}$$

$$r_i = \rho_i/\pi a^2, \tag{8.51}$$

where C_m is membrane capacitance per unit area and ρ_i is the intracellular resistivity. Also, if g_K, g_{Na}, g_l, and I_A are the conductances and applied current density per unit area, then (8.49) becomes the standard form of the first of the set of four Hodgkin–Huxley equations:

$$C_m \frac{\partial V}{\partial t} = \frac{a}{2\rho_i} \frac{\partial^2 V}{\partial x^2} + \bar{g}_K n^4 (V_K - V) + \bar{g}_{Na} m^3 h (V_{Na} - V)$$

$$+ g_l (V_l - V) + I_A, \tag{8.52}$$

$$\frac{\partial n}{\partial t} = \alpha_n (1 - n) - \beta_n n, \tag{8.53}$$

$$\frac{\partial m}{\partial t} = \alpha_m (1 - m) - \beta_m m, \tag{8.54}$$

$$\frac{\partial h}{\partial t} = \alpha_h (1 - h) - \beta_h h. \tag{8.55}$$

This system is often called the *full or complete Hodgkin–Huxley system*. The second space derivative in (8.52) enables the depolarization at one set of space points to initiate changes at neighboring space points. The possibility arises of a local response (cf. solutions of the cable equation), but there is also the possibility of propagating action potentials.

The Hodgkin–Huxley system (8.52)–(8.55) can only be solved by numerical techniques (see Section 8.7). In the original treatment (Hodgkin and Huxley 1952d), the full system was not solved. Instead, the *traveling-wave equations* were solved.

To understand how these arise, suppose at $t = 0$, the spatial distribution of potential is $\bar{V}(x)$ as sketched in Figure 8.11. If this spatial distribution of potential moves to the right with speed u, then at $t = 1$, the solution will be $V(x, 1) = \bar{V}(x - u)$ as the value at x is now that which it was a distance u to the left when $t = 0$. In general, at time t such a traveling-wave solution must be $\bar{V}(x - ut)$.

Now, put

$$z = x - ut, \tag{8.56}$$

so

$$\frac{\partial \bar{V}}{\partial x} = \frac{\partial \bar{V}}{\partial z} \frac{\partial z}{\partial x} = \frac{d\bar{V}}{dz}, \tag{8.57}$$

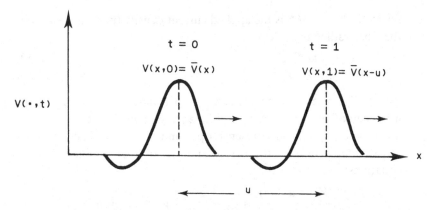

Figure 8.11. Traveling wave of depolarization moving to the right with speed u.

and

$$\frac{\partial \overline{V}}{\partial t} = \frac{\partial \overline{V}}{\partial z}\frac{\partial z}{\partial t} = -u\frac{d\overline{V}}{dz}. \tag{8.58}$$

Since $\overline{V}$ satisfies (8.52), we now have the following system of ordinary differential equations – the *traveling-wave equations*:

$$\frac{d^2\overline{V}}{dz^2} + \frac{2\rho_i C_m u}{a}\frac{d\overline{V}}{dz} = \frac{2\rho_i}{a}\left[\bar{g}_K\bar{n}^4\left(\overline{V} - V_K\right) + \bar{g}_{Na}\bar{m}^3\bar{h}\left(\overline{V} - V_{Na}\right)\right.$$

$$\left. + g_l\left(\overline{V} - V_l\right)\right], \tag{8.59}$$

$$\frac{d\bar{n}}{dz} = \frac{1}{u}\left[\beta_n\bar{n} - \alpha_n(1 - \bar{n})\right], \tag{8.60}$$

$$\frac{d\bar{m}}{dz} = \frac{1}{u}\left[\beta_m\bar{m} - \alpha_m(1 - \bar{m})\right], \tag{8.61}$$

$$\frac{d\bar{h}}{dz} = \frac{1}{u}\left[\beta_h\bar{h} - \alpha_h(1 - \bar{h})\right]. \tag{8.62}$$

This is a five-dimensional autonomous system of first-order equations as can be seen by reintroducing the longitudinal current $I_i = -r_i\,\partial V/\partial x$. Then (8.59) becomes the two equations:

$$\frac{d\overline{V}}{dz} = -\frac{\bar{I}_i}{r_i}, \tag{8.63}$$

$$\frac{d\bar{I}_i}{dz} + \frac{2C_m u \rho_i \bar{I}_i}{a} = \frac{2r_i\rho_i}{a}\left[\bar{g}_K\bar{n}^4\left(\overline{V} - V_K\right) + \bar{g}_{Na}\bar{m}^3\bar{h}\left(\overline{V} - V_{Na}\right)\right.$$

$$\left. + g_l\left(\overline{V} - V_l\right)\right]. \tag{8.64}$$

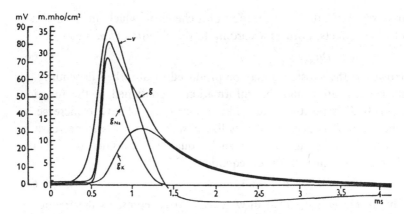

Figure 8.12. Computed traveling-wave solutions of the Hodgkin–Huxley equations. [From Hodgkin and Huxley (1952d). Reproduced with the permission of The Physiological Society and the authors.]

The traveling-wave solutions may be computed numerically, using, for example, the Runge–Kutta method. The speed u of propagation is an unknown parameter. The problem is to pick the correct value of u, which will give solutions that go to the resting state at $z = \pm \infty$. The point $(V, I_i, n, m, h) = (0, 0, 0.35, 0.06, 0.6)$ is the resting *equilibrium point* of the system – at this point all the derivatives are zero. Such points are also called *critical or singular points*. A solution of the kind desired, that is, one which starts and ends at the same equilibrium point, is called a *homoclinic orbit or trajectory* (cf. a *heteroclinic* orbit, which starts at one equilibrium point and ends up at another).

Hodgkin and Huxley (1952d) integrated the traveling-wave equation with a hand calculator. They found that if the speed chosen was too small or too large, then solutions $\overline{V}$ diverged to either $+\infty$ or $-\infty$. The speed calculated was 18.8 m/s at 6.3°C, which compares reasonably well with the experimental speed of 21.2 m/s. Their computed traveling-wave solution (V, g_K, g_{Na}) is shown in Figure 8.12.

8.6 Nonlinear reaction–diffusion systems

The Hodgkin–Huxley equations (8.52)–(8.55) are in the form of a *reaction–diffusion system*. An example of the form of such a system, when there is just one dependent variable $u(x, t)$, is

$$u_t = Du_{xx} + F(u), \tag{8.65}$$

where $D > 0$ is the diffusion coefficient (in square distance per time). The terminology comes from the chemical literature. The quantity u

may represent the concentration of a chemical, which, in the absence of other effects, diffuses according to the *diffusion* (*heat*) equation

$$u_t = Du_{xx}. \tag{8.66}$$

However, the substance may be produced or absorbed depending on its concentration and this information is contained in the function $F(\cdot)$. If F is positive it acts like a *source* as it tends to increase u, whereas if F is negative it acts like a *sink* as it tends to decrease u.

To get some insight into such equations, consider one of the simplest nonlinear diffusion equations

$$u_t = Du_{xx} + u(1 - u). \tag{8.67}$$

This is *Fisher's equation* in which $u \in [0,1]$ represents the frequency of a certain gene in a population that occupies a region in one-dimensional space. If $u = 0$ the gene is totally absent, whereas if $u = 1$ the entire population is of that gene type.

Consider the corresponding *kinetic equation*

$$du/dt = F(u) = u(1 - u), \tag{8.68}$$

which is obtained by setting $u_{xx} = 0$ in (8.67) and is the space-clamped form of Fisher's equation. A stability analysis is straightforward for (8.68). Plotting $F(u)$ versus u (as in Section 3.5 for the nonlinear Lapicque model) gives the graph shown in Figure 8.13. It is seen that if u is initially between 0 and 1, then its derivative is positive and it will increase to asymptotically approach (as $t \to \infty$) the critical point $u = 1$. If u is initially negative, du/dt is negative and $u \to -\infty$ as $t \to \infty$. The directions of the arrows on the u-axis represent the behavior in time of $u(t)$. The origin is seen to be locally *unstable*, whereas $u = 1$ is *asymptotically stable*.

Does this kind of behavior carry over to the partial differential equation (8.67)? In other words, given initial data

$$u(x,0) = f(x), \tag{8.69}$$

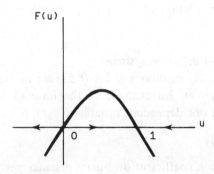

Figure 8.13. Stability analysis (arrow diagram) for Equation (8.68).

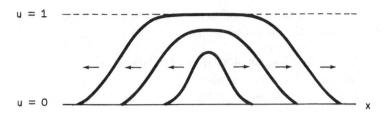

Figure 8.14. Sketch of the formation of a saturating-wave solution of Fisher's equation.

with $f(x) > 0$ for x on some interval (x_1, x_2) of nonzero length, will u grow (or decrease) to the value 1 as $t \to \infty$ for all $x \in (-\infty, \infty)$? That is, do we have

$$\lim_{t \to \infty} u(x, t) = 1, \tag{8.70}$$

or, put another way, is $u = 1$ asymptotically stable and $u = 0$ unstable for (8.67) as it is for (8.68)? The question has been answered for a more general form of Fisher's equation in the affirmative in a theorem of Aronson (1978) called the *hair-trigger effect*. Any positive stimulus leads to $u = 1$ everywhere. Thus a local elevation of u may lead to the sequence of profiles sketched in Figure 8.14. A *saturating wave* may form, which means that the entire population eventually gets taken over by the gene under consideration. The wave is *stable* and its waveform and velocity may be calculated [see Fisher (1937)]. For a general discussion of the properties of solutions, see Murray (1977).

A reaction–diffusion system with two components may take the form

$$u_t = D_1 u_{xx} + F(u, v), \tag{8.71}$$

$$v_t = D_2 v_{xx} + G(u, v), \tag{8.72}$$

with corresponding kinetic equations

$$du/dt = F(u, v), \tag{8.73}$$

$$dv/dt = G(u, v). \tag{8.74}$$

When the dependent variables are coupled in this way, the possibility of traveling-pulse solutions or *solitary waves* arises. These arise in the Hodgkin–Huxley system as well as in models of other neurobiological phenomena such as *spreading cortical depression*. This is a wave that spreads slowly (at speeds of a few millimeters per minute) across the brain as a result of noxious stimuli, involving depolarizations of nerve cells and glia as well as changes in concentrations of various ions and transmitters (Tuckwell and Miura 1978; Tuckwell 1980a, 1981a; Tuckwell and Hermansen 1981). A perturbation of the source function led in that case to *soliton* solutions (Tuckwell 1979b, 1980b); that is, solitary waves that do not annihilate on collision.

8.7 A numerical method of solution of nonlinear reaction–diffusion equations

The form of the reaction–diffusion system of interest is

$$\mathbf{u}_t = \mathbf{D}\mathbf{u}_{xx} + \mathbf{F}(\mathbf{u}), \tag{8.75}$$

where $\mathbf{u} = \mathbf{u}(x, t)$ is the vector $(u_1(x, t), u_2(x, t), \ldots, u_n(x, t))^T$, $\mathbf{D}$ is an $n \times n$ diagonal matrix of diffusion coefficients $D_1, \ldots, D_n$, and $\mathbf{F}(\cdot)$ is a vector-valued function $(F_1(\mathbf{u}), F_2(\mathbf{u}), \ldots, F_n(\mathbf{u}))^T$, T denoting transpose. The Hodgkin–Huxley equations (8.52)–(8.55) can be written in this form with $(V, n, m, h) = (u_1, u_2, u_3, u_4)$ and $D_2 = D_3 = D_4 = 0$. A numerical method for integrating (8.75) has been given by Lees (1967), who modified a scheme called the *Crank–Nicolson method* to allow for the possibly nonlinear reaction term $F(u)$.

Preliminaries: the heat equation

We need a few preliminaries, which can be explained with reference to the heat equation

$$u_t = Du_{xx}, \tag{8.76}$$

where $u = u(x, t)$ is a scalar. A numerical integration procedure is possible by *finite differencing*. In this procedure (8.76) is replaced by a difference equation and the solution of the difference equation approximates that of the partial differential equation.

On the space interval of interest we use $m + 1$ space points with index i and a prescribed number $n + 1$ of time points with index j. Thus, denoting the approximate solution by U,

$$U_{i,j} \simeq u(i\Delta x, j\Delta t), \quad i = 0, 1, \ldots, m; \; j = 0, 1, \ldots, n. \tag{8.77}$$

We assume that the mesh points are equally spaced in the space variable and in the time variable. In the finite-difference approximations, the relevant derivatives are approximated as follows:

$$u_t(x, t) \simeq \frac{U_{i,j+1} - U_{i,j}}{\Delta t}, \tag{8.78}$$

$$u_x(x, t) \simeq \frac{U_{i+1,j} - U_{i,j}}{\Delta x}, \tag{8.79}$$

$$u_{xx}(x, t) \simeq \frac{u_x(x, t) - u_x(x - \Delta x, t)}{\Delta x}$$

$$\simeq \frac{U_{i+1,j} - 2U_{i,j} + U_{i-1,j}}{\Delta x^2}, \tag{8.80}$$

[but see Equation (8.108) for a better approximation for u_x].

To numerically integrate the differential equation by the finite-difference method, a relation is needed between the $U_{i,j+1}$'s and the $U_{i,j}$'s. One method [see Ames (1977)] approximates the second space derivative at t by that at $t + \Delta t$. Then an approximate solution of the heat equation is given by the scheme

$$\frac{U_{i,j+1} - U_{i,j}}{\Delta t} = \frac{D}{\Delta x^2}\left[U_{i+1,j+1} - 2U_{i,j+1} + U_{i-1,j+1}\right].$$

(8.81)

Crank and Nicolson (1947) use the average of the approximations to the second space derivative at the jth and $(j+1)$th time points

$$\frac{U_{i,j+1} - U_{i,j}}{\Delta t} = \frac{D}{2\Delta x^2}\left[U_{i+1,j+1} - 2U_{i,j+1} + U_{i-1,j+1}\right.$$

$$\left. + U_{i+1,j} - 2U_{i,j} + U_{i-1,j}\right].$$

(8.82)

More generally, a weight factor λ can be used with weight λ for the $(j+1)$th time points and weight $(1-\lambda)$ for the jth, with $0 \leq \lambda \leq 1$. Thus, with

$$r = D\Delta t/\Delta x^2,$$

(8.83)

we have

$$-r\lambda U_{i-1,j+1} + (1 + 2r\lambda)U_{i,j+1} - r\lambda U_{i+1,j+1}$$

$$= r(1-\lambda)U_{i-1,j} + [1 - 2r(1-\lambda)]U_{i,j} + r(1-\lambda)U_{i+1,j},$$

(8.84)

where all terms in $j+1$ are collected on the left. Since $i = 0, 1, \ldots, m$, there are $m+1$ equations in $m+1$ unknowns. This integration scheme is called *implicit* because a linear system must be solved to obtain the values of $u(x, t)$ at the next time step. In an *explicit* method, the values at the next time step are immediately obtained in terms of those at the previous time step (cf. the Runge–Kutta method).

The system (8.84) is in *tridiagonal form* and can be solved without matrix inversion as will be seen shortly. If we put

$$X_i = U_{i,j+1},$$

(8.85)

for fixed j, then (8.84) becomes

$$b_0 X_0 + c_0 X_1 = d_0,$$

$$a_i X_{i-1} + b_i X_i + c_i X_{i+1} = d_i, \qquad i = 1, 2, \ldots, m-1,$$

(8.86)

$$a_m X_{m-1} + b_m X_m = d_m,$$

where the coefficients a_i, b_i, c_i, and d_i can be read off from the original system.

The reaction term: Lees' method

We now consider the introduction of the reaction term $F(u)$ into the scalar heat equation

$$u_t = Du_{xx} + F(u). \tag{8.87}$$

In the Crank–Nicolson method the second space derivative is approximated by the average of its finite-difference approximations at time points j and $j+1$. A similar estimate is needed for $F(u)$. Let us temporarily drop reference to the space-variable subscript and set $U_j = U_{i,j}$. With reference to Figure 8.15 we need $U_{j+1/2}$, or actually $F(U_{j+1/2})$.

We have

$$U_{j+1/2} \simeq U_j + \tfrac{1}{2}(U_{j+1} - U_j). \tag{8.88}$$

However, a nonlinear $F(\cdot)$ will make this approximation useless. We therefore approximate the increment in $(j, j+1)$ by that in $(j-1, j)$ and set

$$U_{j+1/2} \simeq U_j + \tfrac{1}{2}(U_j - U_{j-1})$$

$$= \tfrac{3}{2}U_j - \tfrac{1}{2}U_{j-1}. \tag{8.89}$$

Hence Lees' modification of the Crank–Nicolson method gives the tridiagonal system

$$-\frac{r}{2}U_{i-1,j+1} + (1+r)U_{i,j+1} - \frac{r}{2}U_{i+1,j+1}$$

$$= \frac{r}{2}U_{i-1,j} + (1-r)U_{i,j} + \frac{r}{2}U_{i+1,j} + \Delta t\, F\!\left(\frac{3}{2}U_{i,j} - \frac{1}{2}U_{i,j-1}\right). \tag{8.90}$$

Special forms of this system of equations will be discussed under "boundary conditions."

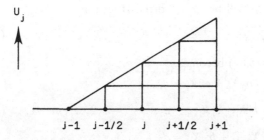

Figure 8.15. The way u is estimated for use in the reaction term $F(u)$ in Lees' method.

It is clear that (8.90) cannot be used for the first time step because when $j = 0$ there appears $U_{i,-1}$, which is not known. *To get started* it is possible to use the explicit formula

$$U_{i,1} = r[U_{i+1,0} - 2U_{i,0} + U_{i+1,0}] + \Delta t\, F(U_{i,0}) + U_{i,0}. \quad (8.91)$$

Application of this formula once, with due regard for the boundary conditions, gives $U_{i,1}$ and $U_{i,0}$ so that formula (8.90) can be used for subsequent time steps.

Solving the tridiagonal system

We write out the tridiagonal system (8.86) to illustrate a method of solution due to Thomas [see Ames (1977), page 52, for additional references]

$$\begin{bmatrix} b_0 & c_0 & 0 & 0 & \cdots & & 0 & 0 \\ a_1 & b_1 & c_1 & 0 & \cdots & & 0 & 0 \\ 0 & a_2 & b_2 & c_2 & \cdots & & 0 & 0 \\ \vdots & \vdots & \vdots & \vdots & & & \vdots & \vdots \\ 0 & 0 & 0 & 0 & \cdots & a_{m-1} & b_{m-1} & c_{m-1} \\ 0 & 0 & 0 & 0 & \cdots & 0 & a_m & b_m \end{bmatrix} \begin{bmatrix} X_0 \\ X_1 \\ X_2 \\ \vdots \\ X_{m-1} \\ X_m \end{bmatrix} = \begin{bmatrix} d_0 \\ d_1 \\ d_2 \\ \vdots \\ d_{m-1} \\ d_m \end{bmatrix}. \quad (8.92)$$

This system is in the form

$$\mathbf{AX} = \mathbf{d}, \quad (8.93)$$

where $\mathbf{A}$ is the $(m + 1) \times (m + 1)$ coefficient matrix, $\mathbf{X} = (X_0, X_1, \ldots, X_m)$, and $\mathbf{d} = (d_0, d_1, \ldots, d_m)$. By performing linear operations (addition of scalar multiples of rows to other rows), it is required to obtain a new coefficient matrix $\mathbf{A}'$ and a new nonhomogeneous vector $\mathbf{d}'$,

$$\mathbf{A}'\mathbf{X} = \mathbf{d}', \quad (8.94)$$

which will make obtaining the solution vector X transparent. It is left as an exercise to show that (8.92) can be transformed to

$$\begin{bmatrix} b_0' & c_0 & 0 & 0 & \cdots & 0 & 0 \\ 0 & b_1' & c_1 & 0 & \cdots & 0 & 0 \\ 0 & 0 & b_2' & c_2 & \cdots & 0 & 0 \\ \vdots & \vdots & \vdots & \vdots & & \vdots & \vdots \\ 0 & 0 & 0 & 0 & \cdots & b_{m-1}' & c_{m-1} \\ 0 & 0 & 0 & 0 & \cdots & 0 & b_m' \end{bmatrix} \begin{bmatrix} X_0 \\ X_1 \\ X_2 \\ \vdots \\ X_{m-1} \\ X_m \end{bmatrix} = \begin{bmatrix} d_0' \\ d_1' \\ d_2' \\ \vdots \\ d_{m-1}' \\ d_m' \end{bmatrix}, \quad (8.95)$$

where

$$b_0' = b_0,$$
$$d_0' = d_0,$$
$$b_i' = b_i - a_i c_{i-1}/b_{i-1}', \qquad i = 1, 2, \ldots, m,$$
$$d_i' = d_i - a_i d_{i-1}'/b_{i-1}', \qquad i = 1, 2, \ldots, m. \tag{8.96}$$

Hence all the coefficients in $\mathbf{A}'$ and the nonhomogeneous terms can be found recursively.

Furthermore, from the last row of (8.95),

$$X_m = d_m'/b_m'. \tag{8.97}$$

The remaining X_i's can be found recursively from

$$X_i = (d_i' - c_i X_{i+1})/b_i', \qquad i = m-1, m-2, \ldots, 1, 0. \tag{8.98}$$

Boundary conditions

Suppose the space interval on which (8.87) is defined is $[0, L]$ and that the initial value of u is given as

$$u(x, 0) = u_0(x), \qquad 0 \le x \le L. \tag{8.99}$$

This gives the values $U_{i,j}$ at $j = 0$ so that the values at $j = 1$ can be found from (8.91) and so can the $j = 2, 3, \ldots$ values by solving the tridiagonal system (8.90).

The general nonhomogeneous linear boundary conditions at $x = 0$ and $x = L$ are

$$k_1 u(0, t) + k_2 u_x(0, t) = k_3, \tag{8.100A}$$
$$k_1' u(L, t) + k_2' u_x(L, t) = k_3'. \tag{8.100B}$$

If $k_2 = 0$ the boundary condition at $x = 0$ is called a *Dirichlet* condition, whereas if $k_1 = 0$ it is referred to as a *Neumann condition*. The details of the tridiagonal system are affected by the nature of the boundary conditions.

Dirichlet conditions at $x = 0$ and $x = L$. For convenience set

$$u(0, t) = \alpha, \tag{8.101A}$$
$$u(L, t) = \beta. \tag{8.101B}$$

In the finite-difference approximation these become

$$U_{0,j} = \alpha, \qquad j = 0, 1, 2, \ldots, \tag{8.102A}$$
$$U_{m,j} = \beta, \qquad j = 0, 1, 2, \ldots. \tag{8.102B}$$

This reduces the number of unknowns in the linear system from

$m + 1$ to $m - 1$. The first of the $m - 1$ equations is, from (8.90),

$$(1 + r)U_{1, j+1} - \frac{r}{2}U_{2, j+1}$$

$$= r\alpha + (1 - r)U_{1, j} + \frac{r}{2}U_{2, j} + \Delta t F\left(\frac{3}{2}U_{1, j} - \frac{1}{2}U_{1, j-1}\right),$$

$$(8.103\text{A})$$

the next $m - 3$ are of the form of (8.90) with $i = 2, \ldots, m - 2$, and the $(m - 1)$th equation is

$$-\frac{r}{2}U_{m-2, j+1} + (1 + r)U_{m-1, j+1}$$

$$= r\beta + \frac{r}{2}U_{m-2, j} + (1-r)U_{m-1, j} + \Delta t F\left(\frac{3}{2}U_{m-1, j} - \frac{1}{2}U_{m-1, j-1}\right).$$

$$(8.103\text{B})$$

The system is still tridiagonal. The starting equation (8.91) becomes, for $i = 1$,

$$U_{1, 1} = r[U_{2, 0} - 2U_{1, 0} + \alpha] + \Delta t F(U_{1, 0}) + U_{1, 0}; \quad (8.104\text{A})$$

it remains the same for $i = 2, 3, \ldots, m - 2$; and for $i = m - 1$ it is

$$U_{m-1, 1} = r[\beta - 2U_{m-1, 0} + U_{m-2, 0}] + \Delta t F(U_{m-1, 0}) + U_{m-1, 0}.$$

$$(8.104\text{B})$$

Neumann conditions at $x = 0$ and $x = L$. Let

$$u_x(0, t) = \alpha, \quad (8.105\text{A})$$

$$u_x(L, t) = \beta. \quad (8.105\text{B})$$

We have seen that a finite-difference approximation for $u_x(x, t)$ is $(U_{i+1, j} - U_{i, j})/\Delta x$. There is a better one obtained as follows. We may write the following Taylor series about x,

$$u(x + \Delta x, t) = u(x, t) + \Delta x\, u_x(x, t) + \Delta x^2\, u_{xx}(x, t)/2 + O(\Delta x^3),$$

$$(8.106\text{A})$$

$$u(x - \Delta x, t) = u(x, t) - \Delta x\, u_x(x, t) + \Delta x^2\, u_{xx}(x, t)/2 + O(\Delta x^3).$$

$$(8.106\text{B})$$

It will be seen that these may be combined to get

$$u_x(x, t) = [u(x + \Delta x, t) - u(x - \Delta x, t)]/2\Delta x + O(\Delta x^2),$$

$$(8.107)$$

which has improved the accuracy by an order of Δx. Accordingly, we should when possible use the *central-difference* approximation

$$u_x(i\Delta x, j\Delta t) \simeq (U_{i+1,j} - U_{i-1,j})/2\Delta x. \qquad (8.108)$$

Then the above Neumann conditions become

$$U_{-1,j} = -2\Delta x\,\alpha + U_{1,j}, \qquad (8.109A)$$

$$U_{m+1,j} = 2\Delta x\,\beta + U_{m-1,j}. \qquad (8.109B)$$

The quantities $U_{-1,j}$ and $U_{m+1,j}$ do not appear above, but we may extend the interval of interest to include them [Ames (1977) calls this the introduction of a false boundary]. The first equation of the tridiagonal system contains $U_{-1,j}$. On substituting (8.109A), the first ($i = 0$) equation becomes, from (8.90),

$$(1+r)U_{0,j+1} - rU_{1,j+1}$$

$$= -2r\Delta x\,\alpha + (1-r)U_{0,j} + rU_{1,j} + \Delta t\,F\left(\frac{3}{2}U_{0,j} - \frac{1}{2}U_{0,j-1}\right), \qquad (8.110A)$$

whereas the starting equation for $i = 0$ becomes

$$U_{0,1} = 2r[U_{1,0} - U_{0,0} - \alpha\,\Delta x] + \Delta t\,F(U_{0,0}) + U_{0,0}. \qquad (8.110B)$$

The equations for $i = 1, 2, \ldots, m-1$ are unaffected. For $i = m$ substitution of (8.109B) in (8.90) gives

$$-rU_{m-1,j+1} + (1+r)U_{m,j+1}$$

$$= 2r\Delta x\,\beta + rU_{m-1,j} + (1-r)U_{m,j} + \Delta t\,F\left(\frac{3}{2}U_{m,i} - \frac{1}{2}U_{m,j-1}\right), \qquad (8.111A)$$

with corresponding starting equation

$$U_{m,1} = 2r[U_{m-1,0} - U_{m,0} + \beta\,\Delta x] + F(U_{m,0}) + U_{m,0}. \qquad (8.111B)$$

The basic system to solve is still tridiagonal.

Mixed boundary conditions at $x = 0$ and $x = L$. If the general boundary conditions (8.100A) and (8.100B) apply with $k_1, k_1' \neq 0$, we may write

$$u(0, t) + \alpha_1 u_x(0, t) = \alpha_2, \qquad (8.112A)$$

$$u(L, t) + \beta_1 u_x(L, t) = \beta_2. \qquad (8.112B)$$

In the finite-difference approximation these become

$$U_{-1,j} = \frac{2\Delta x}{\alpha_1} U_{0,j} + U_{1,j} - \frac{2\alpha_2 \Delta x}{\alpha_1},$$

$$(8.113A)$$

$$U_{m+1,j} = \frac{2\beta_2 \Delta x}{\beta_1} + U_{m-1,j} - \frac{2\Delta x}{\beta_1} U_{m,j}.$$

$$(8.113B)$$

The first equation in the tridiagonal system becomes

$$\left\{ 1 + r\left(1 - \frac{\Delta x}{\alpha_1} \right) \right\} U_{0,j+1} - rU_{1,j+1}$$

$$= -\frac{2\alpha_2 + \Delta x}{\alpha_1} + \left\{ 1 - r\left(1 - \frac{\Delta x}{\alpha_1} \right) \right\} U_{0,j} + rU_{1,j}$$

$$+ \Delta t\, F\left(\frac{3}{2} U_{0,j} - \frac{1}{2} U_{0,j-1} \right),$$

$$(8.114A)$$

whereas the corresponding starting equation is

$$U_{0,1} = 2r\left[U_{1,0} - \left(1 - \frac{\Delta x}{\alpha_1} \right) U_{0,0} - \frac{\alpha_2}{\alpha_1} \Delta x \right] + \Delta t\, F(U_{0,0}).$$

$$(8.114B)$$

The last ($i = m$) equation is

$$-rU_{m-1,j+1} + \left\{ 1 + r\left(1 + \frac{\Delta x}{\beta_1} \right) \right\} U_{m,j+1}$$

$$= \frac{2r\beta_2 \Delta x}{\beta_1} + rU_{m-1,j} + \left\{ 1 - r\left(1 + \frac{\Delta x}{\beta_1} \right) \right\} U_{m,j}$$

$$+ \Delta t\, F\left(\frac{3}{2} U_{m,j} - \frac{1}{2} U_{m,j-1} \right),$$

$$(8.115A)$$

with corresponding starting equation

$$U_{m,1} = 2r\left[U_{m-1,0} - \left(1 + \frac{\Delta x}{\beta_1} \right) U_{m,0} + \frac{\beta_2 \Delta x}{\beta_1} \right] + \Delta t\, F(U_{n,0}).$$

$$(8.115B)$$

When the boundary conditions are not of the same type at $x = 0$ as at $x = L$, the appropriate system of equations can be selected.

Units and standard constants for the squid axon
With reference to Equation (8.52) the usual units are as follows: V, V_K, V_{Na}, and V_l are in millivolts; a and x in centimeters;

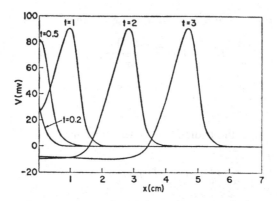

Figure 8.16. Computed solution of the Hodgkin–Huxley equations in response to a current pulse. Voltage is plotted against distance at various times. [From Cooley and Dodge (1966). Reproduced from *The Biophysical Journal* by copyright permission of The Biophysical Society.]

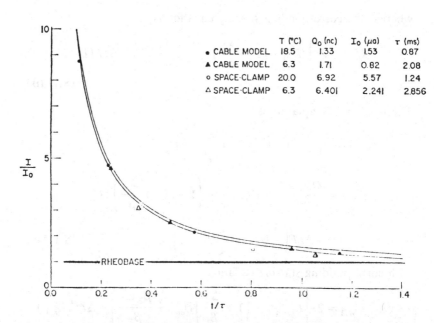

Figure 8.17. Strength–duration curve for the squid axon computed from the Hodgkin–Huxley equations by Cooley and Dodge (1966). (Reproduced from *The Biophysical Journal* by copyright permission of The Biophysical Society.)

in milliseconds; $\bar{g}_K$, $\bar{g}_{Na}$, and $\bar{g}_l$ in millimhos per square centimeter; $I_A(x, t)$ in microamperes per square centimeter; ρ_i, resistivity of axoplasm, in ohm-centimeters; C_m, membrane capacitance per unit area, in microfarads per square centimeter; and L, length of axon, in centimeters.

The following values of the constants for squid axon have been much used and are appropriate for 6.3°C: $a = 0.0238$ cm; $\rho_i = 34.5$ Ω cm; $C_m = 1$ $\mu F/cm^2$; $V_K = -12$ mV; $V_{Na} = 115$ mV; $V_l = 10.613$ mV; $\bar{g}_{Na} = 120$ mΩ^{-1}/cm^2; $\bar{g}_K = 36$ mΩ^{-1}/cm^2; $\bar{g}_l = 0.3$ mΩ^{-1}/cm^2.

The value of $\bar{g}_l$ is chosen to make the rest state stable. g_l is regarded as constant. The α's and β's given above are for 6.3°C. For other temperatures they are scaled with a Q_{10} of 3. That is, they are multiplied by $3^{(T-6.3)/10}$, where T is the temperature in degrees Celsius. Note that the differential equations for n, m, and h may be integrated using the Runge–Kutta method.

Figure 8.16 shows Cooley and Dodge's (1966) computed solution to the full Hodgkin–Huxley equations in response to a current pulse of 10 μA lasting for 0.2 ms. The solitary wave of potential is quickly established. The strength–duration curve computed by the same authors is shown in Figure 8.17.

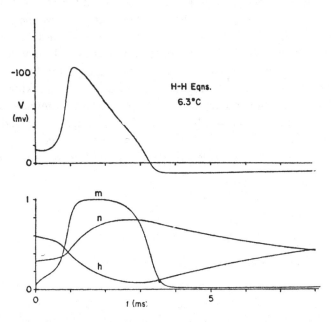

Figure 8.18. Calculated V, m, h, and n as functions of time in the space-clamped Hodgkin–Huxley equations. [From Fitzhugh (1960). Reproduced from *The Biophysical Journal* by copyright permission of The Biophysical Society.]

It is instructive, for the purposes of the approximations we will soon consider, to see how V, m, h, and n vary during the action potential. Figure 8.18 shows this variation for the space-clamped equations.

8.8 The Fitzhugh–Nagumo equations

The analysis of the Hodgkin–Huxley equations (8.52)–(8.55) is extremely difficult because of the nonlinearities and the large number of variables. This is also the case for the space-clamped equations (8.38)–(8.41) and the traveling-wave equations (8.59)–(8.62). Although there are efficient numerical methods of solution of these systems of equations, it is a formidable task to compute solutions with all different sets of parameters, different applied currents, and different boundary conditions of interest. Mathematical analysis would be helpful even if it were performed on simpler equations whose solutions shared the qualitative properties of those of the Hodgkin–Huxley equations. Analysis of such simpler systems may lead to the discovery of new phenomena, which may then be searched for in the original system and also in experimental preparations.

Such a simplified system of equations has its origins in the works of Fitzhugh (1961) and Nagumo, Arimoto, and Yoshizawa (1962) and has become known as the *Fitzhugh–Nagumo equations*. In the Hodgkin–Huxley system the variables V (voltage) and m (sodium activation) have similar (mostly fast) time courses and the variables n (potassium activation) and h (sodium inactivation) have similar (slower) time courses (see Figure 8.18). Heuristically speaking, in the Fitzhugh–Nagumo system, V and m are regarded as mimicked by a single variable $v(x, t)$, which we will call the voltage, and n and h are mimicked by a single variable $w(x, t)$, which is called the *recovery variable*. *The Fitzhugh–Nagumo equations in their general form are* (Rinzel 1979)

$$v_t = v_{xx} + f(v) - w, \tag{8.116}$$

$$w_t = b(v - \gamma w), \tag{8.117}$$

where $f(v)$ is the cubic

$$f(v) = v(1 - v)(v - a), \qquad 0 < a < 1; \tag{8.118}$$

b and γ are positive constants. A term $I = I(x, t)$ representing an applied current may be inserted on the right-hand side of (8.116).

Often γ is set equal to zero in which case we will refer to the *simplified Fitzhugh–Nagumo equations*

$$v_t = v_{xx} + f(v) - w, \tag{8.119}$$

$$w_t = bv, \tag{8.120}$$

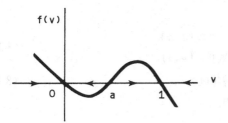

Figure 8.19. The reaction term in the reduced Fitzhugh–Nagumo equation.

which are sometimes combined into the single equation

$$v_t = v_{xx} + f(v) - b\int_0^t v(x, t')\, dt', \tag{8.121}$$

where the last term takes the form of a *killing* term. If we also set $b = 0$, we obtain the *reduced Fitzhugh–Nagumo equation* with just one component

$$v_t = v_{xx} + f(v), \tag{8.122}$$

with kinetic (space-clamped) equation

$$dv/dt = f(v). \tag{8.123}$$

A plot of $f(v)$ versus v appears in Figure 8.19 and shows equilibrium points at $v = 0, a, 1$. A stability analysis (see the directions of the arrows on the v-axis) reveals that the points $v = 0$ and $v = 1$ are (asymptotically) stable and the point $v = a$ is unstable. Hence, if the initial voltage $v(0)$ is between 0 and a, then $v \to 0$ as $t \to \infty$, whereas if $v(0) > a$, $v \to 1$ as $t \to \infty$. The rest point $v = 0$ is therefore locally stable and there is a *threshold effect*. The question arises, as in the previous section, as to whether these stability properties extend to the reduced Fitzhugh–Nagumo equation (8.122).

8.8.1   The reduced Fitzhugh–Nagumo equation

In the reduced system there is no recovery variable. Intuitively, we expect a local elevation of v, corresponding to excitation of the nerve cell, to either decay to rest if too small or give rise to a saturating wave if large enough. We therefore examine threshold effects for (8.122).

Threshold effects

There have been very few studies on threshold effects. Some useful analysis has been performed by Aronson and Weinberger (1975) and Aronson (1978). A more general $f(v)$ has been employed with the properties

$$f(v)\begin{cases} <0, & v \in (0,a), \\ >0, & v \in (a,1), \end{cases} \tag{8.124A}$$

$$f'(0) < 0, \qquad f'(1) > 0, \tag{8.124B}$$

$$f(0) = f(1) = 0, \tag{8.124C}$$

$$\int_0^1 f(v)\, dv > 0. \tag{8.124D}$$

The last of these conditions, when $f(v)$ is the cubic (8.118), is equivalent to

$$a < \tfrac{1}{2}, \tag{8.125}$$

so that the area between the positive part (sources) of f and the v-axis is greater than the negative part (sinks). Conditions (8.124A)–(8.124C) are clearly satisfied by the cubic (8.118) and $v \equiv 0$ and $v \equiv 1$ (i.e., identically zero and one for all x) are equilibrium solutions of (8.122) as they are solutions of the steady-state equation

$$d^2\bar{v}/dx^2 + f(\bar{v}) = 0. \tag{8.126}$$

We are interested in knowing whether an initial distribution of voltage

$$v(x,0) = v_0(x), \qquad -\infty < x < \infty, \tag{8.127}$$

will tend toward one or the other of the equilibrium solutions. Using a comparison theorem, Aronson and Weinberger proceed as follows. Suppose $\chi \in (0,1)$ is such that $\int_0^\chi f(v)\, dv = 0$. Given $\varepsilon \in (\chi, 1)$, there exists a $b_\varepsilon > 0$ and a function $\bar{v}_\varepsilon(x)$, which is a solution of (8.126) satisfying $\bar{v}_\varepsilon(\pm b_\varepsilon) = 0$ with $\bar{v}_\varepsilon(0) = \varepsilon$ and $\bar{v}_\varepsilon(x) \le \varepsilon$ for $x \in (-b_\varepsilon, b_\varepsilon)$. Then the following is true.

Theorem 8.1 (Aronson and Weinberger)
   Let $v(x,t) \in [0,1]$ be a solution of (8.122) with initial condition (8.127). If $f(v)$ satisfies (8.124A)–(8.124D) and

$$v_0(x) \ge \bar{v}_\varepsilon(x), \qquad x \in (-b_\varepsilon, b_\varepsilon),$$

for some $\varepsilon \in (\chi, 1)$, then

$$\lim_{t \to \infty} v(x,t) = 1.$$

The result is depicted in Figure 8.20A. Note that the values of $v_0(x)$ outside $(-b_\varepsilon, b_\varepsilon)$ do not matter; as long as there is enough excitation over a certain finite space interval, v will approach unity everywhere.

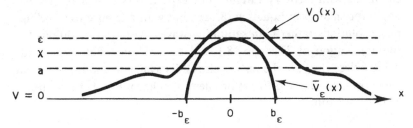

Figure 8.20A. Illustrating the use of Theorem 8.1. The initial voltage displacement $v_0(x)$ is large enough to make $v \to 1$ for all x as $t \to \infty$. For an explanation of the symbols, see the text.

The following result indicates that sufficiently small distributions of excitation will give subthreshold responses, which eventually die away everywhere.

Theorem 8.2 (Aronson and Weinberger)

Let $v(x, t) \in [0, 1]$ be a solution of (8.122) with initial condition (8.127), and let $f(v)$ satisfy (8.124A)–(8.124D). Then for each $\eta \in [0, a)$, there exists a constant $c(\eta) \geq 0$, depending on η, such that if

$$\int_{-\infty}^{\infty} \max\{v_0(x) - \eta, 0\} \, dx \leq c(\eta)(a - \eta),$$

then

$$\lim_{t \to \infty} v(x, t) = 0.$$

The underlying principle is depicted in Figure 8.20B.

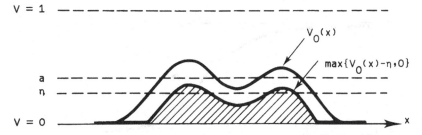

Figure 8.20B. Illustrating the use of Theorem 8.2. If the hatched area is too small, the voltage tends to zero everywhere. For further details, see the text.

Although these threshold results are not for the Fitzhugh–Nagumo equations with recovery variable, one expects that similar results will apply for small b in Equation (8.121), but with reference to traveling-pulse solutions rather than saturating waves. Also, the conditions for the application of Theorems 8.1 and 8.2 are somewhat specialized; nevertheless they are the only such results available and do represent some progress in the difficult problem of delineating what constitutes a threshold stimulus.

Speed of the saturating wave

If there is a saturating-wave solution of the reduced Fitzhugh–Nagumo equation, traveling from left to right with speed u, then we put $v(x, t) = V(x - ut) = V(z)$ as in Section 8.5. Then $V(z)$ satisfies the ordinary differential equation

$$V'' + uV' + f(V) = 0, \tag{8.128}$$

with

$$V(-\infty) = 1, \qquad V(\infty) = 0. \tag{8.129}$$

Solutions thus appear as in Figure 8.21. The speed may be calculated in the following way after Hunter, McNaughton, and Noble (1975). Let $\dot{V} = dV/dz$. Then, since

$$\frac{d^2V}{dz^2} = \frac{d\dot{V}}{dz} = \frac{d\dot{V}}{dV}\frac{dV}{dz} = \frac{d\dot{V}}{dV}\dot{V}, \tag{8.130}$$

Equation (8.128) becomes

$$\frac{d\dot{V}}{dV} = -\left(u + \frac{f(V)}{\dot{V}}\right). \tag{8.131}$$

We must have $\dot{V} = 0$ when $V = 0$ and $V = 1$ and we expect $\dot{V}$ to become negative as V increases from zero, achieve a minimum, and decrease to zero at $V = 1$. A guess is therefore made that the depen-

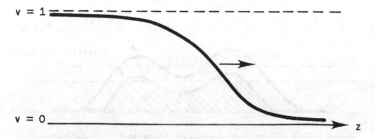

Figure 8.21. Profile of a saturating-wave solution of the reduced Fitzhugh–Nagumo equation.

dence of $\dot{V}$ on V is of the form

$$\dot{V} = kV(V-1),$$ (8.132)

where $k > 0$ is a constant to be determined. From (8.132)

$$d\dot{V}/dV = 2kV - k,$$ (8.133)

and substituting this in the left-hand side of (8.131) gives

$$2kV - k = V/k - u + a/k.$$ (8.134)

For this to be an identity, coefficients of V and unity must be equal on both sides of the equation. This gives

$$2k = 1/k,$$ (8.135A)

$$k = u + a/k.$$ (8.135B)

Solving these gives $k = 1/\sqrt{2}$ and the relation between speed and the parameter a,

$$u = \sqrt{2}\left(\tfrac{1}{2} - a\right).$$ (8.136)

This shows that the speed is a linearly decreasing function of a and that no propagation is possible when $a > \tfrac{1}{2}$ [cf. (8.125)]. The maximum speed occurs at $a = 0$ in which case $f(v) = v^2(1 - v)$. Furthermore (McKean 1970, citing Huxley), the solution of the traveling-wave equation (8.128) and (8.129) with u satisfying (8.136) is

$$V(z) = 1/\left[1 + \exp(z/\sqrt{2})\right]$$ (8.137)

as can be verified as an exercise.

Some physiological insight can be obtained if we adopt the approach of Hunter et al. The cubic source term was inserted as a current–voltage relation in the first Hodgkin–Huxley equation (8.52). This $I–V$ relation is a reasonable "average" representation as it is approximately obtained at certain fixed times after the application of a voltage clamp (Hodgkin et al. 1952). Thus traveling-wave solutions are sought for $V(x, t)$ satisfying

$$C_m \frac{\partial V}{\partial t} = \frac{a}{2\rho_i} \frac{\partial^2 V}{\partial x^2} + gV\left(1 - \frac{V}{V_{th}}\right)\left(1 - \frac{V}{V_p}\right),$$ (8.138)

where g is a conductance per unit area, V_{th} is a threshold voltage (corresponding to the parameter a), and V_p is the peak voltage for the saturating wave (corresponding to $v = 1$). Note that in (8.138), a is the radius of the cylinder! Using the same techniques as above, the speed of the saturating wave is found to be

$$u = \frac{S}{C_m} \sqrt{\frac{ga}{2\rho_i(S+1)}}, \tag{8.139}$$

where

$$S = \frac{V_p}{2V_{\text{th}}} - 1 \tag{8.140}$$

is the *safety factor*. Thus *the speed is proportional to the square root of the axon radius, the square root of the conductance per unit area, and inversely proportional to the capacitance per unit area and the resistivity of the intracellular fluid*. These are useful analytic results because the wave speed is not expected to be very different for small *b* from that of the saturating wave. These also give a useful guide as to what one might expect from solutions of the Hodgkin–Huxley equations and also from real axons.

8.8.2 Space-clamped Fitzhugh–Nagumo equation

As pointed out in Section 8.4, in laboratory preparations the potential of nerve membrane may be held practically uniform in space over a patch of nerve membrane. This is the space-clamped case in which space derivatives are zero. Motivation for using this "lumped-parameter assumption" also comes from the observation that sometimes the solutions of ordinary differential equations, obtained in this way, qualitatively resemble those of the partial differential equations from which they are derived (Conway, Hoff, and Smoller 1978).

The space-clamped Fitzhugh–Nagumo equations obtained from (8.116) and (8.117) are, in the presence of an applied current $I(t)$,

$$dv/dt = f(v) - w + I(t), \tag{8.141}$$

$$dw/dt = b(v - \gamma w). \tag{8.142}$$

We suppose that an impulsive current is applied so $I(t) = Q\,\delta(t)$. This takes v to some value v_0 depending on Q. For $t > 0$, (8.141) applies with $I = 0$ and a qualitative phase-plane analysis is performed as follows.

An *isocline* is a curve in the (v, w)-plane along which the derivative is constant and a *null isocline* is one on which the derivative is zero. It follows that the null isocline for v, on which $dv/dt = 0$, is the curve $w = f(v)$ – the dashed curve in Figure 8.22. Above this curve we have $dv/dt < 0$ and below it $dv/dt > 0$. The null isocline for w is the straight line $w = v/\gamma$, above which $dw/dt < 0$ and below which $dw/dt > 0$. The intersection of these dashed curves gives the *equilibrium*

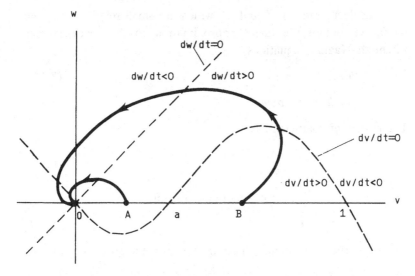

Figure 8.22. Phase-plane analysis for the space-clamped Fitzhugh–Nagumo equations. The dashed lines are those on which $dv/dt = 0$ and $dw/dt = 0$. The only critical point is $(0,0)$, which is asymptotically stable. Responses to voltage steps at $t = 0$ are shown by solid curves with arrows in the direction of increasing time.

(*critical, singular*) points of the system for which such points dv/dt and dw/dt are both zero. The value of γ is chosen so that $(0,0)$ is the only equilibrium point for (8.141) (with $I = 0$) and (8.142). As we will prove shortly, $(0,0)$ *is stable and represents the rest state*.

Consider a trajectory that starts at A with an initial voltage less than a. At A, dv/dt is negative and dw/dt is positive, so the phase point (v, w) moves to the left and upward. It crosses the line at which $dw/dt = 0$ with zero slope. To the left of this line dw/dt is negative so the solution returns to $(0,0)$, crossing the curve $dv/dt = 0$ with infinite slope ($dv/dw = 0$). This trajectory is identified as a *subthreshold response*.

A trajectory starting at B with initial voltage greater than a, must move to the right and upward, making a much larger circuit as shown to eventually return to the origin. This is identified as a *suprathreshold response* and might correspond to a propagating-pulse solution in the nonspace-clamped case.

Stability of the rest point $(0,0)$

If trajectories are to end up at $(0,0)$, the resting point, then when solutions $(v(t), w(t))$ get close to this point they must be drawn into it. We perform a *linearized stability analysis* around $(0,0)$.

If we neglect terms in v^2 and v^3, which are small relative to v near $v = 0$, we obtain the linear approximation to the space-clamped Fitzhugh–Nagumo equations

$$dv/dt = -av - w, \qquad (8.143)$$

$$dw/dt = bv - b\gamma w. \qquad (8.144)$$

The matrix of coefficients is

$$\mathbf{A} = \begin{bmatrix} -a & -1 \\ b & -b\gamma \end{bmatrix}, \qquad (8.145)$$

and the *eigenvalues* λ of $\mathbf{A}$ are obtained from

$$\det(\mathbf{A} - \lambda \mathbf{I}) = 0, \qquad (8.146)$$

where $\mathbf{I}$ is the 2×2 identity matrix. This gives the quadratic equation

$$\lambda^2 + \lambda(a + b\gamma) + b(1 + \gamma a) = 0. \qquad (8.147)$$

The two eigenvalues λ_1 and λ_2 are the roots of this equation

$$\lambda_{1,2} = \frac{-(a + b\gamma) \pm \sqrt{(a + b\gamma)^2 - 4b(1 + \gamma a)}}{2}. \qquad (8.148)$$

From the theory of linear first-order systems, of which (8.143) and (8.144) are an example, solutions may be written as

$$\begin{bmatrix} v(t) \\ w(t) \end{bmatrix} = c_1 \boldsymbol{\xi}_1 e^{\lambda_1 t} + c_2 \boldsymbol{\xi}_2 e^{\lambda_2 t}, \qquad (8.149)$$

where $\boldsymbol{\xi}_1$ and $\boldsymbol{\xi}_2$ are *eigenvectors* of $\mathbf{A}$ with eigenvalues λ_1 and λ_2, respectively, and c_1 and c_2 are arbitrary constants determined by some initial condition. From (8.148) it is seen that the eigenvalues always have negative real parts for the parameter ranges of interest. Hence, no matter what values c_1 and c_2 take, we have, in the linear approximation,

$$\lim_{t \to \infty} \begin{bmatrix} u(t) \\ v(t) \end{bmatrix} = \begin{bmatrix} 0 \\ 0 \end{bmatrix}, \qquad (8.150)$$

so the origin is certainly asymptotically stable for the linear approximation. By a theorem from stability theory (Boyce and DiPrima 1977), the origin is also asymptotically stable for the nonlinear system of space-clamped Fitzhugh–Nagumo equations. In the next section we will see how this situation alters when there is an applied sustained current.

8.8.3 Traveling-pulse solutions

We are interested in the traveling-wave solutions of the simplified Fitzhugh–Nagumo equations (8.119) and (8.120). A proof of the existence of solitary-wave solutions using perturbation techniques was given by Casten, Cohen, and Lagerstrom (1975), but we will sketch a simpler demonstration after Hastings (1975).

If we examine solutions consisting of waves moving from left to right with speed u, then we put $v(x, t) = V(x - ut) = V(s)$ and $w(x, t) = W(x - ut) = W(s)$. This gives the traveling-wave equations

$$V'' = W - uV' - f(V),$$
$$W' = -bV/u. \tag{8.151}$$

To facilitate the analysis we introduce $U = V'$, giving three first-order equations,

$$V' = U,$$
$$U' = -uU - f(V) + W, \tag{8.152}$$
$$W' = -bV/u.$$

For an equilibrium point we must have $V' = U' = W' = 0$. From the first and third equations, we must have $U = V = 0$ at such a point, and, hence, from the second equation, we require $W = 0$ also. Hence $(0, 0, 0)$ is the *only* critical point for (8.152).

We analyze the linear approximation to (8.152) near the critical point in the hope of discovering something about the properties of solutions of (8.152). Linearizing gives

$$\begin{bmatrix} V \\ U \\ W \end{bmatrix} = \begin{bmatrix} 0 & 1 & 0 \\ a & -u & 1 \\ -b/u & 0 & 0 \end{bmatrix} \begin{bmatrix} V \\ U \\ W \end{bmatrix} \tag{8.153}$$

The eigenvalues λ_i of the coefficient matrix are the roots of

$$\lambda^3 = -u\lambda^2 + a\lambda - b/u. \tag{8.154}$$

A graphical solution shows that for small enough b there is one negative eigenvalue λ_1 and two positive eigenvalues λ_2 and λ_3. Solutions of (8.153) may therefore be written as

$$\begin{bmatrix} V(s) \\ U(s) \\ W(s) \end{bmatrix} = c_1 \xi_1 e^{\lambda_1 s} + c_2 \xi_2 e^{\lambda_2 s} + c_3 \xi_3 e^{\lambda_3 s}, \tag{8.155}$$

where ξ_i, $i = 1, 2, 3$, are the eigenvectors of the coefficient matrix and

c_i, $i = 1, 2, 3$, are arbitrary constants determined by some initial condition.

If $c_1 = 0$ and not both c_2 and c_3 are zero, we obtain a two-dimensional family (contained in the plane spanned by the eigenvectors ξ_2 and ξ_3) of solutions that approach the origin as $s \to -\infty$. Similarly, if $c_2 = c_3 = 0$ and $c_1 \neq 0$, a one-dimensional family (along the direction of the eigenvector ξ_1) of solutions is obtained that approaches the origin as $s \to \infty$. These are sketched in Figure 8.23A. Thus $(0, 0, 0)$ is an unstable saddle point for (8.153). By the *stable manifold theorem* [see, for example, Coddington and Levinson (1955), page 330], the solutions of the nonlinear system (8.152) are guaranteed to have similar qualitative features near $(0, 0, 0)$. Thus there is a two-dimensional manifold $\mathcal{M}$ (corresponding to the plane for the linear system) containing $(0, 0, 0)$ such that trajectories enter the origin on $\mathcal{M}$ as $s \to -\infty$. Furthermore, there is a curve $\mathcal{C}$ containing $(0, 0, 0)$ such that solutions approach the origin along $\mathcal{C}$ as $s \to \infty$. This is sketched in Figure 8.23B. Note that $\mathcal{M}$ splits $\mathcal{C}$ into two disjoint branches labeled $\mathcal{C}^+$ and $\mathcal{C}^-$. The question remains as to whether a solution may start on a branch of $\mathcal{C}$ and end up on $\mathcal{M}$. This would give a homoclinic orbit, existing at the rest point at $s = -\infty$, doing a loop in the phase space, and reentering the origin at $s = \infty$, giving a pulse solution as in Figure 8.23C. The right choice for u ensures that this can, in fact, occur if $0 < a < \frac{1}{2}$ and b is small enough. If u is not correct, solutions will go off to either $+\infty$ or $-\infty$, which is what Hodgkin and Huxley found when they numerically integrated the

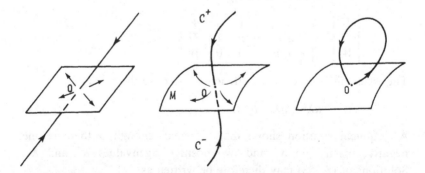

Figure 8.23. A – One-dimensional family of solutions that approach the origin and two-dimensional family of solutions that recede from the origin for the linear system (8.153). B – Representation of the one and two-dimensional manifolds $\mathcal{M}$ and $\mathcal{C}$ of solutions of the nonlinear system. C – Homoclinic orbit representing a traveling-pulse solution.

traveling-wave equation. The following is proved in Hastings (1976a), which should be consulted for further results.

If $0 < a < \frac{1}{2}$ *and b is small enough, there are two distinct positive numbers* u_* *and* u^* *such that (8.152) has homoclinic solutions.*

Thus there are traveling-pulse solutions of the simplified Fitzhugh–Nagumo equations. However, this is more than we asked for as there are two wave speeds for a given set of parameter values. A speed diagram is shown in Figure 8.24 [after McKean (1970)]. For each $a \in (0, \frac{1}{2})$ there are two wave speeds for each value of b. Note that for $u = 0$ and $b = 0$ an explicit solution is available (Cohen 1971)

$$V(s) = \frac{2a\left[\left(ke^{-\sqrt{a}s} + \dfrac{1+a}{3}\right) - \dfrac{1+a}{3}\right]}{\left(ke^{-\sqrt{a}s} + \dfrac{1+a}{3} - \dfrac{\sqrt{2a}}{2}\right)\left(ke^{-\sqrt{a}s} + \dfrac{1+a}{3} + \dfrac{\sqrt{2a}}{2}\right)},$$

$$k = \frac{a}{V(0)} - \frac{1+a}{3}. \tag{8.156}$$

This gives the part of the speed diagram on the a-axis.

Numerical calculations of solutions of the simplified Fitzhugh–Nagumo equations [see McKean (1970)] indicate that the wave with the larger speed, for given a and b, is stable. By stability is meant

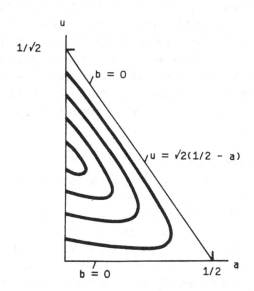

Figure 8.24. Speed diagram for the simplified Fitzhugh–Nagumo equations. The wave speed u is plotted against a for various values of b. The upper and lower edges of the triangle correspond to $b = 0$.

here that small disturbances in the waveform will disappear as the wave propagates. Analytical proofs of this are available for modified Fitzhugh–Nagumo equations in which $f(v)$ is replaced by a piecewise linear approximation as suggested by McKean (1970). Rinzel and Keller (1973) and Rinzel (1975) have proved the existence of two pulse solutions when f has the form

$$f(v) = H(v - a) - v, \qquad 0 < a \leq \tfrac{1}{2}, \tag{8.157}$$

and found the one with the lower speed to be unstable [see also Feroe (1978)]. Proofs of the stability of the fast traveling-pulse solutions of the Fitzhugh–Nagumo equations have been given by Jones (1984) and Yanagida (1985).

8.9 Responses to sustained inputs: repetitive activity

Sufficient depolarization or exciting current applied locally in space and/or time to a nerve cell may elicit an action potential. When such inputs are sustained, the possibility arises of trains of several, and possibly infinite numbers (theoretically), of action potentials. However, the question as to when this will occur is not so easily answered. In this work we are only considering nerve cells in *isolation*. Review at this point Figure 1.19, which shows the complex *network* of interconnected cells in which a motoneuron participates. It is clear that when one delivers (naturally or experimentally) a sustained input to such a nerve cell, there may be several subsequent inputs due to the synaptic excitation (with interacting inhibition) of other cells in the network that may exert feedback influences on the cell in question. Other complications involve shifts in ionic concentrations, which in turn affect membrane potentials and conductances for various ions. However, we will find that even for a cell in isolation, the matter of repetitive activity is a complex one.

8.9.1 Subthreshold oscillations

Subthreshold responses of the space-clamped Hodgkin–Huxley model and of the squid axon when a voltage step is applied were seen in Figure 8.10. The membrane potential decays and passes to hyperpolarized levels to eventually return to resting level. The larger the initial voltage displacement, the greater the level of subsequent hyperpolarization. Sabah and Leibovic (1969) solved the full Hodgkin–Huxley system numerically for sustained subthreshold current steps at $x = 0$ and found the kind of behavior shown in Figure 8.25. The potential undergoes a damped oscillation to achieve a steady-state value in about 10 ms. Larger subthreshold current steps

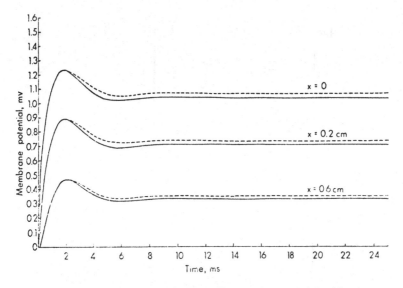

Figure 8.25. Subthreshold oscillations in potential with constant current (0.2 μA) at $x = 0$ obtained by numerical solution of the Hodgkin–Huxley equations. The temperature is 20°C. Dashed curves represent the cable model. [From Sabah and Leibovic (1969). Reproduced from *The Biophysical Journal* by copyright permission of The Biophysical Society.]

lead to greater amplitude oscillations. Sabah and Leibovic were able to obtain reasonable agreement between such solutions of the Hodgkin–Huxley equations and a linearized version of them when the current was small. It should be noted that oscillations do not occur in all nerve cells – for example, the spinal motoneuron (Figure 3.3) and the crab axon (Figure 5.4). Linear cable theory (Chapters 4–6) does not predict oscillatory responses to steady currents.

8.9.2 Repetitive activity with application of a constant current

When discussing repetitive activity we must be careful to distinguish theoretical results obtained by solving (usually) the Hodgkin–Huxley equations and experimental results. Another important distinction must be made between the space-clamped and nonspace-clamped case, the latter receiving our attention first because it is more natural.

Numerical solution of the Hodgkin–Huxley equations (Cooley, Dodge, and Cohen 1965; Cooley and Dodge 1966; Stein 1967b) reveals repetitive firing overlimited ranges of a constant applied current. Figure 8.26 is taken from Stein (1967b) and shows the computed voltages at the point of application of the current ($x = 0$)

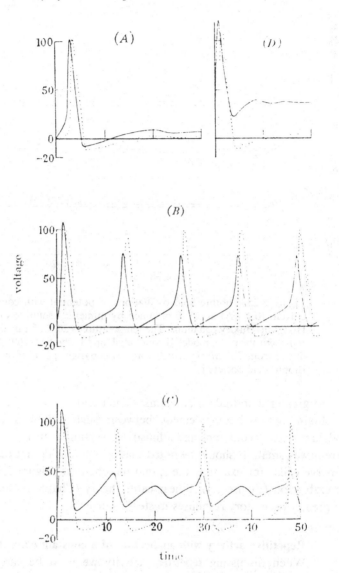

Figure 8.26. Responses of the Hodgkin–Huxley system to sustained constant currents. The solid curve is the response at $x = 0$, the dotted one at a distance 1 cm from the point of application of the current. For A–D the currents are 0.76, 2.53, 6.1, and 10 μA, respectively. [From Stein (1967b). Reproduced with the permission of The Royal Society and the author.]

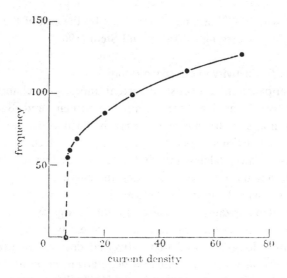

Figure 8.27. Frequency of repetitive firing versus input current as obtained from the Hodgkin–Huxley equations. [From Stein (1967b). Reproduced with the permission of The Royal Society and the author.]

and at a distance of 1 cm (dotted curves). In Figure 8.26A a weak current step elicits only one action potential, whereas in Figure 8.26B a stronger current elicits a train of action potentials. In Figure 8.26C repetitive firing is still present but the response near the stimulus is quite different from that further along the axon. In Figure 8.26D only one action potential is seen.

Thus there is a range of constant current intensities over which repetitive firing occurs. At the upper end of this range there is a sudden drop in frequency to zero, a phenomenon called *nerve block*. The frequency-versus-current relation obtained by Stein is given in Figure 8.27. Note that there is a discontinuity in the f/I curve when a sudden jump in frequency of repetitive firing occurs from 0 to about 50 impulses/s at currents of about 1 μA. Also, the amplitude of the oscillations decreases near the point of application of the current as the current strength increases in the range of currents that give repetitive firing.

In the squid axon repetitive firing is rare, at least in laboratory preparations. Hagiwara and Oomura (1958) could only elicit a few action potentials from the squid axon with constant applied currents. The situation is quite different in other preparations. Crayfish stretch receptors may fire repetitively for several minutes under constant input current with a 30% decline in frequency by the end of the train (*accommodation*), according to Brown and Stein (1966).

8.9.3 Repetitive activity under space clamp

The application of a constant current under space clamp may induce repetitive firing [see, for example, Guttman and Barnhill (1970)]. It is not easy to induce but may be facilitated by lowering the external *calcium* concentration from its normal values. Frankenhaeuser and Hodgkin (1957) found that reducing external calcium levels has the effect of translating the curve of peak sodium conductance versus voltage in the depolarizing direction. This means that a smaller depolarization is required to induce a given increase in sodium conductance so the cell becomes more easily excited. Frankenhaeuser and Hodgkin found that a fivefold decrease of external Ca^{2+} concentration was equivalent to a depolarization of 10–15 mV.

Calculations on the space-clamped Hodgkin–Huxley equations [see, for example, Stein (1967b), Guttman and Barnhill (1970), Sabah and Spangler (1970), Rinzel (1978), Holden (1980), and Rinzel and Miller (1980)] indicate that higher frequencies of repetitive activity are possible under space clamp. Figure 8.28A shows voltage trajectories so obtained for two values of the applied constant current and Figure 8.28B gives the variation of frequency and amplitude of the oscillations as a function of the applied current. It is seen that there is repetitive activity in a well-defined range of current intensitives. As I increases in this range, the frequency increases monotonically, whereas the amplitude of the oscillations decreases after an initial increase. Holden (1976a) computed the response of the space-clamped Hodgkin–Huxley system to cyclic input currents. Phase locking of output to input was found similar to that in the Lapicque model (c.f. Section 3.8)

8.9.4 Repetitive activity in the space-clamped Fitzhugh–Nagumo system

Analysis of the Hodgkin–Huxley equations is facilitated by first examining the simpler Fitzhugh–Nagumo equations, which have been more energetically studied. Under space clamp and a constant applied current I, these equations are

$$\frac{dv}{dt} = f(v) - w + I \doteq F_1(v, w),$$

(8.158)

$$\frac{dw}{dt} = b(v - \gamma w) \doteq F_2(v, w).$$

(8.159)

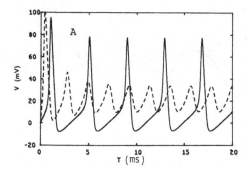

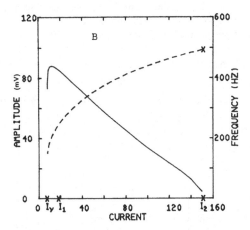

Figure 8.28. A – Computed voltage trajectories for the space-clamped Hodgkin–Huxley equations for current steps of $I = 10$ $\mu A/cm^2$ (solid curve) and $I = 12$ $\mu A/cm^2$ (dashed curve) at 18.5°C. B – Amplitude and frequency (dashed curve) of oscillatory responses. [From Rinzel (1978). Reproduced with the permission of Federation Proceedings and the author.]

When $\gamma = 0$ the only critical point is $(v, w) = (0, I)$ and it will be seen that this point is always stable. Also, there are no periodic solutions that would correspond to repetitive activity (McKean 1970).

When $\gamma \neq 0$ there may be three critical points for small enough I and certain parameter values. However, for certain parameter ranges, there is only one critical point for each value of I. The phase-plane picture is then as sketched in Figure 8.29, where the curves on which $dv/dt = 0$ and $dw/dt = 0$ are shown intersecting at (v_c, w_c). From (8.159) we must have $w_c = v_c/\gamma$ so v_c is the unique solution of

$$f(v_c) - v_c/\gamma + I = 0. \tag{8.160}$$

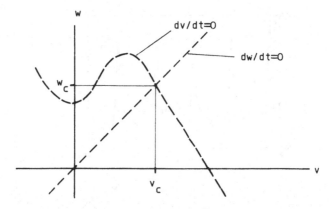

Figure 8.29. The phase-plane picture for the space-clamped Fitzhugh–Nagumo equations with applied current in the case where there is only one singular point (v_c, w_c).

To obtain insight into the solution properties a stability analysis is performed on the system obtained from (8.158) and (8.159) by linearizing about the critical point. The coefficient matrix $\mathbf{A}$ of the linear approximation ($\bar{v}(t)$, $\bar{w}(t)$),

$$\frac{d}{dt}\begin{bmatrix}\bar{v}\\\bar{w}\end{bmatrix} = \mathbf{A}\begin{bmatrix}\bar{v}\\\bar{w}\end{bmatrix},$$

(8.161)

is

$$\mathbf{A} = \begin{bmatrix}\dfrac{\partial F_1}{\partial v} & \dfrac{\partial F_1}{\partial w}\\[2mm]\dfrac{\partial F_2}{\partial v} & \dfrac{\partial F_2}{\partial w}\end{bmatrix}.$$

(8.162)

This gives

$$\mathbf{A} = \begin{bmatrix}f'(v_c) & -1\\b & -b\gamma\end{bmatrix}.$$

(8.163)

The eigenvalues of $\mathbf{A}$, which determine the stability properties of the linear approximation, are the roots of

$$\det(\mathbf{A} - \lambda\mathbf{I}) = 0,$$

(8.164)

which gives

$$\lambda^2 + \lambda(b\gamma - f'(v_c)) + b(1 - \gamma f'(v_c)) = 0.$$

(8.165)

The eigenvalues are thus

$$\lambda_{1,2} = \frac{f'(v_c) - b\gamma \pm \sqrt{(b\gamma - f'(v_c))^2 - 4b(1 - \gamma f'(v_c))}}{2}.$$

(8.166)

We look for changes in the stability properties of the critical point of the linear system as the applied current changes. These are likely to occur when the eigenvalues cross the imaginary axis. From (8.166) we see that the eigenvalues will be *pure* imaginary if the following two conditions are met:

$$f'(v_c) = b\gamma, \tag{8.167}$$

$$4b(1 - \gamma f'(v_c)) > (b\gamma - f'(v_c))^2. \tag{8.168}$$

Putting (8.167) in (8.168) gives the further simple condition

$$b\gamma^2 < 1. \tag{8.169}$$

Using the definition of $f(\cdot)$ in (8.67) gives the quadratic equation

$$3v_c^2 - 2v_c(1 + a) + (a - b\gamma) = 0, \tag{8.170}$$

with roots

$$v_{c_{1,2}} = \frac{(1 + a) \pm \sqrt{(1 + a)^2 - 3(a - b\gamma)}}{3}, \tag{8.171}$$

which are both positive. Hence there are two distinct positive values of v_c for which the coefficient matrix $\mathbf{A}$ has pure imaginary eigenvalues. Also, v_c is a strictly increasing function of I. Thus there are two distinct positive values of I, which we designate I_1 and I_2, with $I_1 < I_2$, at which $\mathbf{A}$ has pure imaginary eigenvalues. The values of I_1 and I_2 are obtained from the equation defining v_c [Equation (8.158) with $dv/dt = 0$];

$$I_i = v_{c_i}\left[v_{c_i}^2 - v_{c_i}(1 + a) + (a + 1/\gamma)\right], \qquad i = 1, 2. \tag{8.172}$$

Further analysis [see Troy (1976) and Rinzel (1979)] shows that for $I < I_1$ or $I > I_2$ the eigenvalues of $\mathbf{A}$ have negative real parts. Thus the critical point is then asymptotically stable, being either a node or spiral point. These results carry over to the nonlinear system (8.158) and (8.159).

Bifurcation theory

When $I = I_1$ the eigenvalues of $\mathbf{A}$ are pure imaginary, and for $I \in (I_1, I_2)$ the eigenvalues of $\mathbf{A}$ have positive real parts so the critical point is unstable. We use I as a parameter; then the critical point changes from asymptotically stable to unstable as I increases through I_1 and similarly as I decreases from above I_2 to less than I_2. This kind of phenomenon, where a change in the stability properties of a critical point occurs as some parameter passes through some particular value, is called a *bifurcation*. The solutions for $I \in (I_1, I_2)$ are completely different in nature from those for $I \notin (I_1, I_2)$. The quantity I is referred to as a *bifurcation parameter*.

To ascertain what happens to solution properties as I passes through the values I_1 and I_2 we appeal to the *Hopf bifurcation theorem*, which has numerous applications in studies of diverse systems (Marsden and McCracken 1976). The following version is adopted from Schmidt (1976), some of the technical points being omitted. For an excellent introductory treatment, see Segel (1980), page 683 et seq.

Let $\mathbf{x}(t) = (x_1(t), x_2(t), \ldots, x_n(t))^T$ satisfy the system of first-order equations

$$d\mathbf{x}/dt = \mathbf{F}(\mathbf{x}(t); \mu), \tag{8.173}$$

where μ is a parameter [that is, $\dot{x}_i = F_i(x_1, x_2, \ldots, x_n; \mu)$, $i = 1, \ldots, n$)]. Equilibrium points are obtained from the condition $\dot{x}_1 = \dot{x}_2 = \cdots = \dot{x}_n = 0$ and we suppose that for each μ there is an equilibrium point $\tilde{\mathbf{x}}(\mu)$ such that

$$\mathbf{F}(\tilde{\mathbf{x}}(\mu); \mu) = \mathbf{0}. \tag{8.174}$$

The Jacobian matrix of first partial derivatives evaluated at the critical point is

$$\mathbf{A}(\mu) = \begin{bmatrix} \dfrac{\partial F_1}{\partial x_1} & \dfrac{\partial F_1}{\partial x_2} & \cdots & \dfrac{\partial F_1}{\partial x_n} \\ \dfrac{\partial F_2}{\partial x_1} & \dfrac{\partial F_2}{\partial x_2} & \cdots & \dfrac{\partial F_2}{\partial x_n} \\ \vdots & \vdots & & \vdots \\ \dfrac{\partial F_n}{\partial x_1} & \dfrac{\partial F_n}{\partial x_2} & \cdots & \dfrac{\partial F_n}{\partial x_n} \end{bmatrix}_{\tilde{\mathbf{x}}(\mu)}, \tag{8.175}$$

and the eigenvalues $\lambda_1, \lambda_2, \ldots, \lambda_n$ of $\mathbf{A}(\mu)$ are the roots of the equation

$$\det(\mathbf{A}(\mu) - \lambda \mathbf{I}) = 0, \tag{8.176}$$

where $\mathbf{I}$ is the $n \times n$ identity matrix. Suppose in a neighborhood of a particular value μ_0 of the parameter μ there is a pair of eigenvalues of $\mathbf{A}(\mu)$ of the form $\alpha(\mu) \pm i\beta(\mu)$ such that $\alpha(\mu_0) = 0$, $\beta(\mu_0) \neq 0$, no other eigenvalue of $\mathbf{A}(\mu_0)$ being an integral multiple of $i\beta(\mu_0)$. Thus $\mathbf{A}(\mu_0)$ has a pair of pure imaginary eigenvalues. Suppose further that the rate of change of the real part, $\alpha'(\mu_0) \neq 0$. Then the following holds.

Theorem 8.3 (Hopf)

Given the above conditions, there exists a one-parameter family, indexed by ε, of real nonconstant periodic solutions of (8.173), $x = x(t; \varepsilon)$, with $\mu = \mu(\varepsilon)$, such that $\mu(0) = \mu_0$, $x(t; 0) = \tilde{x}(\mu_0)$, and the period $T(\varepsilon)$ satisfies $T(0) = 2\pi/\beta(\mu_0)$.

For the space-clamped Fitzhugh–Nagumo system, the current I plays the role of the bifurcation parameter μ, and the values I_1 and I_2 in turn play the role of μ_0. The system of interest (8.158) and (8.159) has the Jacobian matrix at the critical point $(v_c, w_c) \equiv (\tilde{x}_1, \tilde{x}_2)$ given by (8.163). Further analysis is required to see on which side of I_1 and I_2 the stationary point bifurcates to the periodic orbits representing repetitive firing and whether the periodic orbits are stable. Such analysis has been performed by Troy (1976) and the following are valid in the parameter ranges given in that article, which should be consulted for further details.

Theorem 8.4 (Troy)

From each of the steady-state solutions $(v_c(I_1), w_c(I_1))$ and $(v_c(I_2), w_c(I_2))$, there is a bifurcation of small periodic solutions of (8.158) and (8.159). Furthermore, there exist $\gamma_1, \gamma_2 > 0$ such that bifurcation from $(v_c(I_1), w_c(I_1))$ occurs on $(I_1, I_1 + \gamma_1)$ and the bifurcation from $(v_c(I_2), w_c(I_2))$ occurs on $(I_2 - \gamma_2, I_2)$. The periodic orbits bifurcating from the steady-state solutions at I_1 and I_2 are asymptotically stable for $I \in (I_1, I_1 + \gamma_1)$ and $I \in (I_2 - \gamma_2, I_2)$.

In Figure 8.30 the phase portraits and voltage-versus-time curves are sketched according to the above analysis. In Figure 8.30A the critical point is stable and solutions are drawn into it as $t \to \infty$. In Figure 8.30B the critical point is unstable but there appears a periodic orbit (closed curve in the phase plane), which is stable. Solutions are attracted to this orbit and oscillations persist indefinitely. This is the case for a larger current (Figure 8.30C) except the amplitude of the oscillations has grown. In Figure 8.30D the amplitude has gotten smaller and in Figure 8.30E the periodic orbit has disappeared with a return to a stable critical point.

For the *nonspace-clamped* simplified Fitzhugh–Nagumo equations (8.119) and (8.120), Hastings (1976a) has demonstrated the existence of periodic solutions for speeds u satisfying $u_* < u < u^*$, where u_*

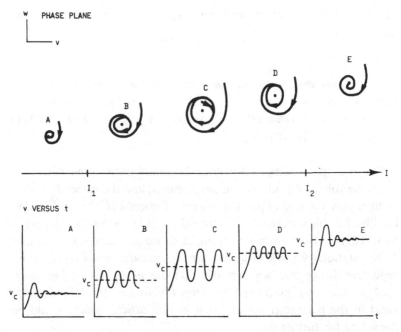

Figure 8.30. Bifurcation to periodic orbits in the space-clamped Fitzhugh–Nagumo equations. For an explanation, see the text.

and u^* are the speeds of the unstable and stable traveling-pulse solutions. More recently, Rinzel and Keener (1983) have applied bifurcation analysis to the solutions of the Fitzhugh–Nagumo system (8.116)–(8.118) and found a Hopf bifurcation to periodic solutions at a critical input current, extending the results of Rinzel (1978) for the case of a piecewise linear approximation to $f(v)$.

8.10   Analyses of the Hodgkin–Huxley equations

The intimidating complexities of the Hodgkin–Huxley equations precluded their analysis for nearly 20 years after their formulation. Although such analysis has been carried far in the last decade, it is not yet completed. The pioneering step was made by Fitzhugh (1960) who applied phase-space techniques to reduced forms of the Hodgkin–Huxley equations under space clamp. An example is the following, where only two variables are studied.

Since V and m move on a fast time scale relative to n and h, it is useful to first consider the equations under the assumptions that

$$\dot{h} = \dot{n} = 0. \tag{8.177}$$

That is, h and n are constants that may be set at their resting values, which we designate h_0 and n_0. This gives the reduced system (V, m), which, in the absence of applied currents, is

$$C\,dV/dt = \bar{g}_K n_0(V_K - V) + \bar{g}_{Na} m^3 h_0(V_{Na} - V) + g_l(V_l - V),$$
(8.178)

$$dm/dt = \alpha_m(V)(1 - m) - \beta_m(V)m.$$
(8.179)

Since now there are just two variables, phase-plane analysis is easily performed. The phase picture is sketched in Figure 8.31. The m and V null isoclines intersect at A, B, and C to give *three critical points*. Linearized stability analysis about these points shows that A and C are *asymptotically stable*, whereas B is an *unstable saddle point*. A is identified as the resting point of the reduced system, whereas C is another stable equilibrium point near the sodium Nernst potential. The saddle point at B is characterized by two incoming and two outgoing trajectories, other trajectories shying away from B should they approach it, as indicated in the figure.

A small current impulse, which takes the system to the point A', results in an immediate return to the rest point. A larger shock, which takes the voltage to the more depolarized level at A'', results in a trajectory, which eventually approaches C. Recall that there is no activation of the potassium conductance and inactivation of the sodium conductance in this reduced system. That is, there is no possibility of recovery from large enough depolarizations. Tasaki and Hagiwara (1957) discovered evidence of the upper equilibrium point C when they experimentally excited squid axon treated with TEA (see

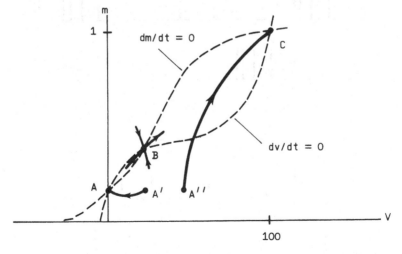

Figure 8.31. Phase portrait for the (V, m)-reduced system. [Adapted from Fitzhugh (1960).]

Section 8.2). The voltage lingered at large depolarizations because the potassium activation had been blocked, but, because there was still sodium inactivation, the voltage eventually returned to rest after a long pause, thus producing an action potential with a large plateau. For further studies of the reduced systems, see Fitzhugh (1960).

The behavior of the (V, m)-reduced system dominates the rise of potential during the action potential and therefore the leading edge of the propagating wave. Hence a study of the (V, m) system, with the diffusion term in the equation for V, is helpful in finding propagation speeds. Scott (1975, 1977) has looked at the (V, m) system in another way by considering the two-component system consisting of the axial current I_i and the voltage V. In the traveling-wave equations for (V, I_i) there are three critical points, two being saddle points and the other a node. The saturating wave, which approximates the leading edge, corresponds to a heteroclinic orbit from one saddle point to the other.

The full Hodgkin–Huxley system (8.52)–(8.55) has been subject to analysis by only a few authors. Evans and Schenk (1970) and Chen (1976) have established the *existence and uniqueness* of the solution to the initial-value problem. Evans (1972a–c, 1975) and Evans and Feroe (1977) have established the stability of the rest point and the stability of the traveling-pulse solutions.

Carpenter (1977, 1978, 1981) has studied traveling-wave solutions in systems of generalized Hodgkin–Huxley equations. Using the *iso-*

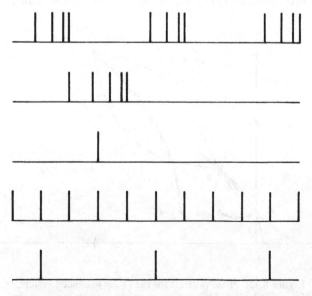

Figure 8.32. Types of solution of the Hodgkin–Huxley equations shown to exist by Carpenter (1981).

lating-blocks method, she was able to establish the existence of homoclinic solitary and multiple-pulse solutions [see also Hastings (1976b)] as well as plateau-type solutions. Such solutions were shown *not* to exist for certain parameter ranges. The *structural stability* [see Cronin (1977)] of the equations under changes in their functional form has thus been at least partially investigated. Several solution types were predicted by Carpenter (1981), which had previously been thought not to exist for the Hodgkin–Huxley equations without the addition of a auxiliary variables and mechanisms. Some of these solutions are sketched in Figure 8.32.

The space-clamped Hodgkin–Huxley equations

The space-clamped Hodgkin–Huxley equations under application of a constant current have been analyzed analytically [see, for example, Troy (1976, 1978)] and numerically [see, for example, Rinzel (1978), Hassard (1978), and Rinzel and Miller (1980)]. Troy has used Hopf bifurcation theory to establish that the qualitative features of the solutions as the current varies are not dissimilar to those of the simpler Fitzhugh–Nagumo equations. For each input current I there is a unique steady-state solution. Bifurcation to periodic solutions occurs at two distinct values of I.

The numerical calculations lend support to these results and have led to the following picture. There exist three current intensities, $I_\nu < I_1 < I_2$, depending on temperature, such that the following oc-

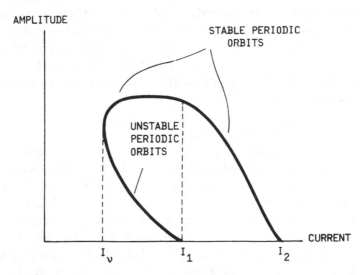

Figure 8.33. Amplitude of oscillatory solutions versus input-current intensity for the space-clamped Hodgkin–Huxley equations. [Adapted from Troy (1978) and Rinzel (1978).]

curs. For $I \notin (I_1, I_2)$ the steady state is stable, whereas for $I \in (I_1, I_2)$ the steady state is unstable. For $I < I_\nu$ there are damped oscillatory solutions, which decay to the steady state. For $I_\nu < I < I_1$ both the steady-state *and* periodic solutions are stable. Depending on how I is tuned into the interval (I_ν, I_1), we may see a steady state prevail or an oscillation. However, between I_ν and I_1 there are also *unstable* periodic solutions that cannot lead to observed repetitive activity. At I_1 and I_2 bifurcation theory predicts the emergence of oscillatory solutions. If we plot amplitude of the oscillatory response versus I, we obtain the picture of Figure 8.33. Holden and Yoda (1981) report that K + channel density, or $\bar{g}_K$, can also be a bifurcation parameter for the space-clamped Hodgkin–Huxley equations. See Holden (1981) for a review of the analyses of the various nerve-conduction models. See also Hodgson (1983) for some recent developments.

8.11 Action potentials in myelinated nerve: the Frankenhaeuser–Huxley equations

As pointed out in Section 1.8, the axon of, for example, the spinal motoneuron is ensheathed for the most part with insulating myelin with breaks at the *nodes* of Ranvier. A typical distance between nodes is of order 1 mm. The myelin sheath has been found to be passive and its properties may be explained in terms of spatially distributed resistance and capacitance both with fixed magnitudes (Huxley and Stampfli 1949). That is, the electrical properties of the myelin sheath are adequately accounted for by linear cable theory (Chapters 4 and 5). The nodal membrane, on the other hand, is active in the sense that it has voltage-dependent conductances in the manner of the squid axon. We first look at the equations that have been developed for nodal voltage and permeability changes before proceeding to examine the propagation of the action potential along the myelinated nerve fiber.

Nodal action potential

A long sequence of experiments by Frankenhaeuser and his co-workers (Dodge and Frankenhaeuser 1958; Frankenhaeuser 1959; Frankenhaeuser and Waltman 1959; Dodge and Frankenhaeuser 1959; Frankenhaeuser 1960; Frankenhaeuser 1962a–c; Frankenhaeuser 1963a–b; Frankenhaeuser and Moore 1963a, b) culminated in the formulation of a system of equations for nodal membrane by Frankenhaeuser and Huxley (1964). Although the underlying principles are similar to those postulated for the squid axon, the details are quite different. The nodal length is small enough that diffusive effects may be ignored. The node may therefore be regarded as a uniformly polarized patch of membrane requiring only a system of ordinary, rather than partial, differential equations for the description of its membrane dynamical processes.

Based on voltage-clamp data for nodal membrane of the toad (and frog) myelinated nerve, the ionic current was divided into four components:

$$I_{\text{ion}} = I_{\text{K}} + I_{\text{Na}} + I_l + I_p, \tag{8.180}$$

which differs from (8.3) in that there is an extra term I_p representing a nonspecific delayed current.

As we saw in Section 2.15, the instantaneous current–voltage relation in nodal membrane does not validate the description of the state of the membrane in terms of the conductances g_{K} and g_{Na}. Rather the constant-field theory expressions for these ionic currents (Section 2.9) were found to provide an excellent fit to the data. From (2.85) we have

$$I_{\text{Na}} = \frac{\gamma F P_{\text{Na}} V_m \left[C_{\text{Na}}^o - C_{\text{Na}}^i e^{\gamma V_m} \right]}{\left(e^{\gamma V_m} - 1 \right)}, \tag{8.181}$$

$$I_{\text{K}} = \frac{\gamma F P_{\text{K}} V_m \left[C_{\text{K}}^o - C_{\text{K}}^i e^{\gamma V_m} \right]}{\left(e^{\gamma V_m} - 1 \right)}, \tag{8.182}$$

where V_m is the actual membrane potential [Equation (2.1)], γ is given by (2.35), and P_{Na} and P_{K} are the sodium and potassium *permeabilities* of the membrane. For consistency, a similar expression was developed for I_p, which, being mainly a sodium current, is written (Frankenhaeuser 1963a) as

$$I_p = \frac{\gamma F P_p V_m \left[C_{\text{Na}}^o - C_{\text{Na}}^i e^{\gamma V_m} \right]}{\left(e^{\gamma V_m} - 1 \right)}. \tag{8.183}$$

The leakage current takes the same form as for squid axon, that is,

$$I_l = g_l (V - V_l), \tag{8.184}$$

where V is the depolarization.

The equations for the sodium permeability are

$$P_{\text{Na}} = \overline{P}_{\text{Na}} m^2 h, \tag{8.185}$$

where $\overline{P}_{\text{Na}}$ is constant, m and h are dimensionless variables, taking values in $[0, 1]$, representing sodium activation and inactivation, and satisfying

$$dm/dt = \alpha_m (1 - m) - \beta_m m, \tag{8.186}$$

$$dh/dt = \alpha_h (1 - h) - \beta_h h. \tag{8.187}$$

Table 8.1. *Constants in the rate coefficients*

i	A_i (ms^{-1})	B_i (mV)	C_i (mV)
1	0.36	22	3
2	0.4	13	20
3	0.1	-10	6
4	4.5	45	10
5	0.02	35	10
6	0.05	10	10
7	0.006	40	10
8	0.09	-25	20

Here the α's and β's are voltage-dependent coefficients

$$\alpha_m = \frac{A_1(V - B_1)}{1 - \exp\{(B_1 - V)/C_1\}}, \tag{8.188}$$

$$\beta_m = \frac{A_2(B_2 - V)}{1 - \exp\{(V - B_2)/C_2\}}, \tag{8.189}$$

$$\alpha_h = \frac{A_3(B_3 - V)}{1 - \exp\{(V - B_3)/C_3\}}, \tag{8.190}$$

$$\beta_h = \frac{A_4}{1 + \exp\{(B_4 - V)/C_4\}}. \tag{8.191}$$

The values of the A's, B's, and C's are given in Table 8.1.

The equation for the potassium permeability is

$$P_K = \bar{P}_K n^2 k, \tag{8.192}$$

where n and k are activation and inactivation variables. However, the time course of the potassium inactivation is so slow that, for studying single action potentials, k may be regarded as constant. Then

$$P_K = P_K' n^2, \tag{8.193}$$

where n satisfies

$$dn/dt = \alpha_n(1 - n) - \beta_n n, \tag{8.194}$$

and

$$\alpha_n = \frac{A_5(V - B_5)}{1 - \exp\{(B_5 - V)/C_5\}}, \tag{8.195}$$

$$\beta_n = \frac{A_6(B_6 - V)}{1 - \exp\{(V - B_6)/C_6\}}. \tag{8.196}$$

Similarly,

$$P_p = \overline{P}_p p^2, \tag{8.197}$$

$$\frac{dp}{dt} = \alpha_p(1-p) - \beta_p p, \tag{8.198}$$

$$\alpha_p = \frac{A_7(V - B_7)}{1 - \exp\{(B_7 - V)/C_7\}}, \tag{8.199}$$

$$\beta_p = \frac{A_8(B_8 - V)}{1 - \exp\{(V - B_8)/C_8\}}. \tag{8.200}$$

In conjunction with the equation

$$I_{\text{ion}} + C\,dV/dt = I_A, \tag{8.201}$$

where I_A is the applied current density (in amperes per square centimeter), Equations (8.180)–(8.200) constitute the *Frankenhaeuser–Huxley equations*.

The standard data are as follows. The constants A_i, B_i, and C_i are given in Table 8.1. The remaining standard data are: $C_{\text{Na}}^o = 114.5$ mM; $C_{\text{Na}}^i = 13.74$ mM; $C_K^o = 2.5$ mM; $C_K^i = 120$ mM; $g_l = 30.3$ mΩ^{-1}/cm^2; $V_l = -0.026$ mV; $C = 2$ μF/cm^2; $\overline{P}_{\text{Na}} = 8 \times 10^{-3}$ cm/s; $P_K' = 1.2 \times 10^{-3}$ cm/s; $\overline{P}_p = 0.54 \times 10^{-3}$ cm/s; $V_R = -70$ mV.

Using a Runge–Kutta method of integration, Frankenhaeuser and Huxley computed the response of the system to current of strength 1 mA/cm^2 applied for 0.12 ms. The resulting action potential is shown in Figure 8.34 along with experimentally obtained action potentials from nodal membrane. The agreement between predicted and experimental voltage trajectories is excellent. However, recent experiments and calculations indicate that the Frankenhaeuser–Huxley equations give an inadequate description of sodium currents. Bromm, Schwarz, and Ochs (1981) treated nodal membrane with TEA to abolish potassium currents and found the duration of the measured action potential was about 2.5 times as long as that predicted by the Frankenhaeuser–Huxley equations.

Propagation of an action potential along a myelinated fiber
In one of the first theoretical studies of conduction in myelinated nerve, Rushton (1951) used a scaling argument to predict that the *speed of the impulse should be proportional to the diameter of the fiber* (where "diameter" means including the myelin sheath). Indeed, experimental evidence supported this idea, in distinction to

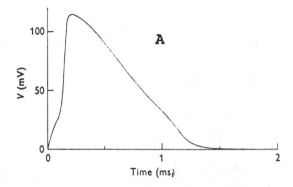

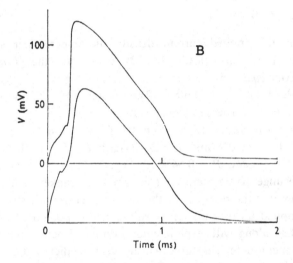

Figure 8.34. A – Action potential for nodal membrane computed from the Frankenhaeuser–Huxley equations. B – Experimentally obtained action potentials. [From Frankenhaeuser and Huxley (1964). Reproduced with the permission of The Physiological Society and the authors.]

the situation with unmyelinated nerve, where the speed was approximately proportional to the *square root* of the axon diameter (cf. Section 8.8.1). Rushton also argued that there must be an optimal thickness for the myelin. If it were too thin, the membrane resistance would be too low and there would be excessive current leaking out between nodes. If it were too thick, there would be a loss of space for the intracellular fluid so the axial resistance would be too high.

Figure 8.35 depicts a myelinated nerve fiber. The point $x = 0$ is chosen arbitrarily. The axon radius is a, the *internodal* distance is L,

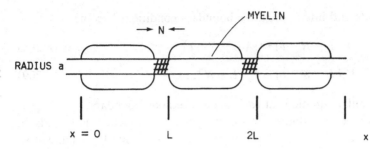

Figure 8.35. Depiction of myelinated nerve.

and the length of the nodes is N. To formulate model equations suitable for investigating propagation, we must consider nodal and internodal regions separately.

On the assumption that the internodes behave like passive cables, we have

$$r_I c_I V_t = (r_I/r_i) V_{xx} - V, \qquad 0 < x < L - N, \tag{8.202}$$

where r_I is the internodal membrane resistance times unit length, c_I is the capacitance per unit length between nodes, and r_i is the axial resistance per unit length. Similar equations will apply at other internodes. The boundary condition at $x = 0$ must be specified according to the experimental conditions.

If we assume that the nodal membrane is *active*, then

$$C_m V_t = (a/2\rho_i) V_{xx} + I_{Na} + I_K + I_p + I_L, \qquad L - N < x < L, \tag{8.203}$$

where C_m is the capacitance per unit area, ρ_i is the intracellular resistivity, and I_{Na}, I_K, I_p, and I_l are given by (8.181)–(8.184). The auxiliary equations are

$$\partial m/\partial t = \alpha_m (1 - m) - \beta_m m, \tag{8.204}$$

$$\partial h/\partial t = \alpha_h (1 - h) - \beta_h h, \tag{8.205}$$

$$\partial n/\partial t = \alpha_n (1 - n) - \beta_n n, \tag{8.206}$$

$$\partial p/\partial t = \alpha_p (1 - p) - \beta_p p. \tag{8.207}$$

Alternatively, the Hodgkin–Huxley form of the expression for the total ionic current may be employed.

The boundary conditions at the interface of internode and node are those demanded by continuity of potential and conservation of axial current. Assuming the internal resistance per unit length is the same for node and internode, these boundary conditions become

$$V(L - N^-, t) = V(L - N^+, t), \tag{8.208}$$

$$V_x(L - N^-, t) = V_x(L - N^+, t), \tag{8.209}$$

with similar equations at each node–internode boundary.

In numerical studies of the propagation of action potentials in myelinated fibers, the nodal membrane has usually been lumped so that its potential satisfies an ordinary, rather than partial, differential equation. The reason for this is apparently that the internodal distance L is usually of order 1000 to 2000 μm, which is much greater than the nodal length N of about 3 μm. However, there is no good reason for this as was alluded to in Section 6.6. A calculation based on the data provided by Brill et al. (1977) gives a space constant for internodal membrane of about 1.2 cm, so that the electrotonic length of the internode is only about 0.1. Furthermore, if we calculate a space constant for nodal membrane based on the maximal value of $g_{Na} + g_K + g_l$ in the Hodgkin–Huxley framework, we obtain a value of about 0.4 μm. In terms of these space constants the nodal membrane is actually longer than the internode. Thus with suitable scaling the lengths of the node and internode may be made about the same so that the same increment in space variable can be employed in the numerical integration of (8.202) and (8.203). The second boundary condition (8.209) would have to be changed accordingly.

There have been several interesting studies on the theory of spike propagation in myelinated nerve [see, for example, Waxman (1977), Brill et al. (1977), and Moore et al. (1978)]. All of these studies employed Hodgkin–Huxley dynamics for nodal membrane. These studies are particularly relevant to understanding the effects of *demyelination*, which occurs in certain diseases such as *multiple sclerosis* and in painful disorders such as *trigeminal neuralgia* (Calvin, Howe, and Loeser 1977). As Waxman (1977) points out, a theoretical study of the effects of demyelination may assist in the suggestion of clinical treatment.

In one such study the effects of a missing myelin sheath between two nodes were investigated. The result is shown in Figure 8.36. It can be seen that the demyelination causes conduction block. This calculation assumed that the demyelinated region had the same properties as *nodal* membrane. On the other hand, if the *entire* nerve is demyelinated, an action potential propagates with a speed about one-tenth of that when the myelin is present. Another complication due to demyelination is that there may be temporal dispersion of a nerve

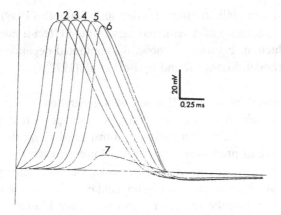

Figure 8.36. Action potentials at nodes 1–7 when the internode between nodes 7 and 8 is stripped of myelin. Calculated results from Waxman (1977). (Reproduced with the permission of The American Medical Association and the author.)

signal when it propagates down a nerve tract (containing several axons) as it propagates more slowly in demyelinated fibers.

Moore et al. (1978) investigated the effects of several parameters on propagation in myelinated nerve. They found that the speed of propagation was insensitive to changes in nodal area and nodal membrane dynamics. Also, the speed was independent of internodal distance over a wide range. There is, nevertheless, a peak velocity in the range 1000 to 2000 μm, which is the range in which most actual

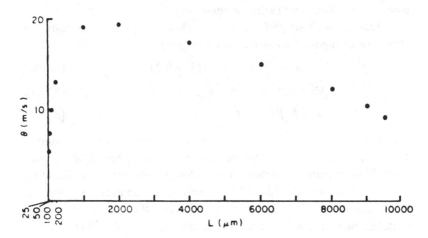

Figure 8.37. Calculated speeds of conduction of action potentials in myelinated nerve for various internodal distances. There is conduction block around 10,000 μm. [From Brill et al. (1977). Reproduced with the permission of *The British Medical Journal* and the authors.]

internodal distances fall. In fact, Huxley and Stampfli (1949) suggested that the geometry of myelinated nerve would effect a maximal speed of conduction. Figure 8.37 shows the computed dependence of speed on internodal distance found by Brill et al. (1977).

8.12 Anatomical factors in action-potential generation
In the idealized picture an action potential is generated at one part of a nerve cell and is then faithfully transmitted along the axon to telodendrites and presynaptic terminals. However, even the axons of real nerve cells are not perfect cylinders as they undergo changes in diameter, course through changing extracellular ionic environments, and interact with neighboring fibers, and they may branch several times either to give axon collaterals or preterminal branches. When one considers the entire cell with its dendritic tree, geometric complications become even more apparent. We will briefly consider changes in diameter, branching, K^+ accumulation in the extracellular space, temperature effects, neurons with dendritic trees, and nerve bundles.

Changes in diameter
Changes in axon (or possibly dendrite) diameter may be either sudden or smooth. The phenomena of interest include *conduction block* and the possibility of *reflected spikes*. There have been several such studies [see, for example, Khodorov et al. (1969), Khodorov (1974), Goldstein and Rall (1974), Parnas, Hochstein, and Parnas (1976), and see Khodorov and Timin (1975) and Swadlow, Kocsis, and Waxman (1980) for reviews].

Goldstein and Rall (1974) used the following equations to study the effects of changes of diameter on propagation:

$$V_t = V_{xx} - V + E(1 - V) - I(V + 0.1), \qquad (8.210)$$

$$E_t = k_1 V^2 + k_2 V^4 - k_3 E - k_4 I, \qquad (8.211)$$

$$I_t = k_5 E + k_6 EI - k_7 I, \qquad (8.212)$$

where $E = E(x, t)$, $I = I(x, t)$, and $k_1, \ldots, k_7$ are positive constants. These equations have solitary-pulse solutions with properties similar in qualitative features to those of the Hodgkin–Huxley equations and are easier to work with. Using numerical methods of solution, Goldstein and Rall investigated the effects of the various geometries sketched in Figure 8.38. Their results are summarized as follows:

   (a) *Propagation into a sealed end.* As the action potential approaches the sealed end, it attains a larger amplitude and greater speed.

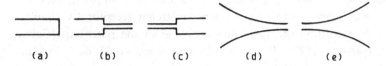

Figure 8.38. Geometries investigated by Goldstein and Rall (1974).

(b) *Sudden decrease in diameter*. Propagation is always successful and as expected the speed is smaller on the smaller diameter cylinder.

(c) *Sudden increase in diameter*. If the ratio of the diameters is not too large the action potential may successfully pass into the larger cylinder. Too large an increase in diameter results in failure to propagate. For some intermediate cases there is not only a propagating action potential in the larger cylinder, but there emerges a backward traveling spike in the smaller cylinder – a reflected spike.

(d) *Smooth decrease in diameter*. Propagation is successful.

(e) *Smooth increase in diameter*. Propagation will occur providing the rate of increase of the diameter is not too large.

Some of these findings are supported by the experimental and theoretical studies on the squid axon by Ramon, Joyner, and Moore (1975). In the experiments axial resistance was changed to mimic a change in diameter. The computations were performed on the Hodgkin–Huxley equations. For diameter ratios greater than three, reflected spikes were observed in the simulations.

The key to understanding whether conduction will be a success or failure at regions of changing diameter is the observation that for propagation there must be sufficient longitudinal current (diffusion) ahead of the wave of depolarization to generate enough membrane current to locally exceed the threshold. This is always possible when the diameter is decreasing, but not when it is increasing.

The phenomenon of reflected spikes at junctions, where there is an increase in diameter, has important implications. In some cells it is established (for example, the spinal motoneuron) that there is a low-threshold region (trigger zone) in the initial part of the axon (initial segment in myelinated nerve). An action potential may form first in the trigger zone and propagate orthodromically along the axon and antidromically into the soma and dendrites. The increase in diameter from axon to some may be sufficient to give rise to a reflected spike. Thus, due to this *retrograde invasion* of the soma, a doublet spike may form. This phenomenon, an example of *reexcitation*, was demonstrated in the lobster stretch receptor by simultaneously recording the soma and axon membrane potentials (Calvin and Hartline 1977). It was found that there could be a somatic recording of a single action potential, whereas a doublet propagated along the axon. This indicates that somatic recording may not always be a reliable guide to firing frequency. Calvin and Hartline also implicated the extra spikes elicited by reexcitation as candidates for certain pathological nerve-cell conditions.

Branching

When an action potential encounters a branch point, there may be complete transmission failure or there may be transmission into one or both daughter cylinders. The failure, or success, of an action potential to pass a branch point may be the basis of the success, or failure, of synaptic response discussed in Section 1.5. Goldstein and Rall (1974) found in their computations that there was either propagation into both daughter cylinders or neither. Propagation through branch points has been studied experimentally and theoretically (using Hodgkin–Huxley equations) for the squid axon by Westerfield, Joyner, and Moore (1978). In the experiments conduction through a branch point was secured and then the temperature increased until conduction failure occurred.

From cable theory (Chapter 4), the input resistance of a semifinite cylinder is

$$R_{in} = \lambda r_i = k/d^{3/2}. \qquad (8.213)$$

If we label the parent 0 and the daughter cylinders 1 and 2, then the combined input resistance of the daughter cylinders is given by

$$1/R = 1/R_1 + 1/R_2, \qquad (8.214)$$

which gives

$$R = k/\left(d_1^{3/2} + d_2^{3/2}\right). \qquad (8.215)$$

The ratio of the input conductance of the prebranch cylinder to the combined input conductance of the postbranch cylinders is thus

$$R_0/R = \left(d_1^{3/2} + d_2^{3/2}\right)/d_0^{3/2}. \qquad (8.216)$$

Success, or failure, of propagation is found to depend on this ratio (Parnas and Segev 1979; Scott and Vota-Pinardi 1982). If the ratio is less than or equal to unity, propagation is successful. As the ratio increases a critical value is reached at which conduction failure occurs. According to both theory and experiment, this critical value is a linearly decreasing function of temperature (Westerfield et al. 1978).

An interesting example of the effects of axonal branching was found by Parnas [see Swadlow et al. (1980)] in the axons of a crustacean. Invasion of the branch point led to successful propagation down one branch when the frequency of impulses was between 40 to 50 s^{-1} but along the other when the frequency was between 80 to 100 s^{-1}.

Extracellular space and K$^+$ accumulation

We have seen that during the action potential potassium ions are extruded. If the extracellular compartment is small and the extracellular concentration of K$^+$ small, as is usually the case, then the extruded K$^+$ may lead to a substantial increase in the external concentration of this ion. This will alter not only the potassium Nernst potential but also the resting membrane potential according to the Goldman–Hodgkin–Katz formula.

Adelman and Fitzhugh (1975) modified the space-clamped Hodgkin–Huxley equations by introducing an extra equation that governed the potassium concentration in the periaxonal space between axon and Schwann cell layer. In addition to allowing V_K to vary according to the Nernst formula, these authors also incorporated changes in $\bar{g}_K$ and the α's and β's in the auxiliary equations.

The inclusion of the effects of varying the external K$^+$ concentration was highly successful as better agreement with the experimental action potential was obtained. Also, when the squid axon is repetitively *stimulated*, a train of action potentials ensues and better agreement was obtained with the features of the spike train. Furthermore, when the modified equations were used to predict the effects of a constant current step, the results again agreed with experiment as now only a few spikes at most were seen compared with an infinite number with the unmodified Hodgkin–Huxley equations. See also Lebovitz (1970).

Parnas et al. (1976) investigated both the effects of accumulating extracellular K$^+$ and the changing diameter of the axon on propagation of action potentials in the giant axon of the cockroach. Experimentally, it was found that a single spike could propagate through the region of changing diameter but a train of impulses was blocked – a phenomenon called *frequency block*. By solving the Hodgkin–Huxley equations modified by the inclusion of variable K$^+$ and allowing the axon diameter to slowly increase, Parnas et al.

successfully demonstrated the failure of propagation of an action-potential train.

Models of neurons with dendritic trees

The first attempt to model action-potential generation in a "complete" nerve cell was that of Dodge and Cooley (1973). They devised a model of a spinal motoneuron using the full Hodgkin–Huxley equations to describe active membrane and employed Rall's equivalent-cylinder concept to treat the dendrites. The components of their model are shown in Figure 8.39. The rate constants and other parameters in the Hodgkin–Huxley equations were adapted to motoneuron data where available. The voltage-clamp data of Araki and Terzuolo (1962) on motoneurons were fitted with the adjusted parameter values. The values of $\bar{g}_K$, $\bar{g}_{Na}$, and $\bar{g}_l$ varied with location throughout the cell. It was found that the best fit with the experimental and computed action potentials was obtained when the following

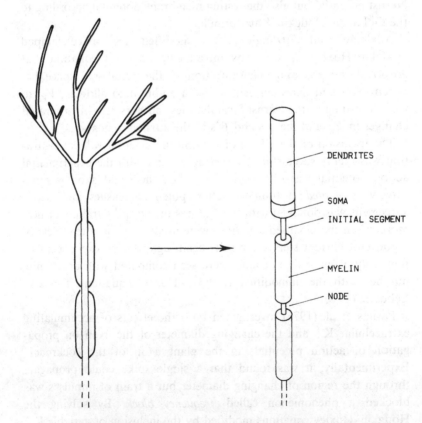

Figure 8.39. Components of the model motoneuron of Dodge and Cooley (1973).

values were employed [but see Hafner et al. (1981) for the application
of nonlinear least-squares methods to this kind of problem] for the
conductances (in mhos per square centimeter):

Region	$\bar{g}_{Na}$	$\bar{g}_K$	g_l
Dendrites	0	0	0.167
Soma	70	17.5	0.167
Initial segment	600	100	1
Myelin	0	0	0.05
Node	600	100	3

It was deduced that the dendrites were completely *passive*, that the
soma was not as active as the initial segment, and that the latter had a
threshold some 10 mV less than the soma.

Traub (1977a) pursued the modeling of the motoneuron further.
The *Dodge–Cooley model* did not predict repetitive firing with constant current injected into the soma, in disagreement with the experimental results (see Chapter 1). It was noted that the lumped-circuit
model of Kernell and Sjöholm (1972, 1973) gave the correct f/I
relations due to the inclusion of slow potassium conductance. This
extra conductance was included with the addition of another auxiliary
equation. Traub divided the motoneuron into seven compartments:
node, first internode, initial segment, soma, and three compartments
representing the various dendritic regions. The soma and initial segment contained active membrane but the slow potassium conductance
was restricted to the soma. Successful prediction of the *primary and
secondary ranges* of repetitive firing was obtained with this model as
well as a lower threshold for repetitive firing for smaller motoneurons
in accordance with the size principle. It was postulated that differences in slow potassium conductance were responsible for the
phasic and *tonic* responses of different motoneurons (see Chapter 1).

The same model was applied in simplified form to Renshaw cells
with a successful prediction of its firing characteristics (Traub 1977b).
Subsequently, the model of the motoneuron was extended to include
calcium conductances that activated the slow potassium conductance
(Traub and Llinás 1977). The slow potassium conductance was found
to control firing frequency in the primary range, whereas the fast
potassium conductance controlled this quantity in the secondary
range. Models have also been developed for pyramidal cells of the
hippocampus (Traub and Llinás 1979) and neocortex (Traub 1979)
that exhibit the *paroxysmal depolarizing shifts* found during *epilepsy*.
A similar model for cortical pyramidal cells has been proposed that

includes active dendritic spines (Shepherd et al. 1985). It has been postulated that enhanced propagation of voltage changes induced by synaptic input can occur because spine–spine excitations occur in a way similar to the saltatory conduction of spikes in a myelinated nerve.

Propagation in nerve bundles

When several nerve axons form a nerve tract, there is the possibility of electrical interactions between individual fibers through the extracellular space. This was demonstrated by Katz and Schmidt (1940) who found that when a pulse in one fiber was followed by a faster pulse in an adjacent fiber, the second pulse would catch up with the first and the two pulses would travel together in a locked-in fashion.

There have been a few theoretical investigations of such interactions. Scott and Luzader (1979) and Eilbeck, Luzader, and Scott (1981) employed the coupled Fitzhugh–Nagumo-type equations

$$v_{1,t} = (1 - \alpha)v_{1,xx} - \alpha v_{2,xx} + f(v_1) - w_1, \tag{8.217}$$

$$w_{1,t} = b(v_1 - \gamma w_1), \tag{8.218}$$

$$v_{2,t} = (1 - \alpha)v_{2,xx} - \alpha v_{1,xx} + f(v_2) - w_2, \tag{8.219}$$

$$w_{2,t} = b(v_2 - \gamma w_2), \tag{8.220}$$

where $\alpha \ll 1$ is a coupling parameter. The solutions of the traveling-wave equations were expanded as a perturbation series in powers of α and an expression obtained for the first-order term in the correction to the speed of propagation when $f(\cdot)$ was piecewise linear.

Bell (1981) has considered a pair of coupled Hodgkin–Huxley axons and provided conditions for the existence of a traveling-pulse solution on both fibers. A result was also obtained in the case of $n > 2$ fibers and for other dynamical equations representing the behavior of active membrane (Bell and Cook 1978, 1979).

9

The stochastic activity of neurons

9.1   Introduction

All of the models of nerve-cell activity we have considered thus far have been *deterministic*. They have consisted of a differential equation (with a threshold condition imposed) or a system of differential equations in conjunction with a given input current and given boundary–initial-value conditions.

To illustrate, consider the linear Lapicque model of Chapter 3. If $V(t)$ is the depolarization at time t, C is the neuron capacitance, R is the neuron resistance, and $I(t)$ is the input current, then

$$C\,dV/dt + V/R = I(t), \qquad t > 0, V < \theta, V(0) = 0, \qquad (9.1)$$

with a spike generated when V reaches or exceeds θ. If $I(t)$ is constant, then the predicted time between action potentials is always the same and is given by (3.52). If $I(t) = I_0 + I_1\cos(\omega t + \phi)$, representing a cyclic input, the time interval between spikes is *variable* (see Figure 3.21) but *the sequence of times of occurrence of spikes is completely and uniquely determined*.

Such deterministic models are inadequate for the description of the activity of real neurons. Assuming for now that the mathematical model (differential equations, boundary conditions, threshold condition, if necessary) is valid, the input current is rarely, if ever, known with certainty. This is true even in controlled experiments such as the current-injection experiments described in Chapter 1 and analyzed in Chapter 8. Even in those experiments the time between action potentials is variable.

In most experimental studies of the activity of nerve cells in more natural conditions, very little, if anything, is known about the inputs to the cell under study. Indeed, one of the challenges has been to ascertain the nature of the input even in relatively simple situations (see Section 6.5). Certain sensory inputs may be controlled but still the actual inputs to the nerve cells are not known. The observed quantity is the train of spikes, or possibly the time course, of the intracellular or extracellular potential, not the input currents, and so forth.

Although action potentials are not instantaneous, it is customary to assign them occurrence times, which may, for example, be the time at which the voltage apparently attains threshold.

Definition

Let $\{\Theta_k, k = 0, 1, 2, \ldots\}$ be a sequence of times at which a nerve cell emits action potentials, with $\Theta_0 = 0$ and $\Theta_0 < \Theta_1 < \Theta_2 < \cdots$. The kth *interspike interval* (ISI) is

$$T_k = \Theta_k - \Theta_{k-1}, \qquad k = 1, 2, \ldots . \tag{9.2}$$

The first measurements that revealed the variability of the ISI were on the muscle spindles of frogs (Brink, Bronk, and Larrabee 1946; Buller, Nicholls, and Strom 1953; Hagiwara 1954). Under conditions of constant tension of the muscle, the ISIs were quite variable and this variability was greatest when the muscle was in its unstretched state. Some results from one such experiment are shown in Figure 9.1, where t_k, the value of T_k, is plotted against k. It can be seen that there is a tendency for the interspike intervals to become longer and longer (*adaptation*), but there is also a haphazard fluctuation.

Spontaneous activity

The term *spontaneous activity* with reference to neural activity is not well defined. It may be loosely described as the activity of a

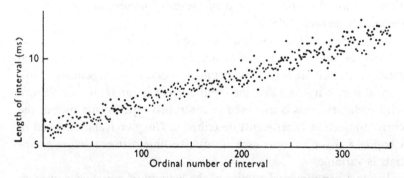

Figure 9.1. Length of interspike intervals versus order of appearance for a muscle spindle held at fixed tension. [From Buller et al. (1953). Reproduced with the permission of The Physiological Society and the authors.]

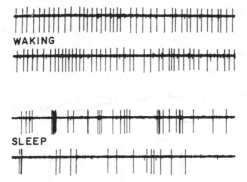

Figure 9.2. Spike trains from a pyramidal tract cell of a monkey while the animal was awake (upper two traces) and asleep (lower two traces). [From Evarts (1964). Reproduced with the permission of The American Physiological Society and the author.]

neuron, with focus on the collection of action potentials it emits, in the absence of an intended input or when there is no obvious input from other cells. The majority of, but not all, CNS neurons exhibit spontaneous activity consisting of an irregular train of impulses. Examples are shown in Figure 9.2, where the spiking activity of a pyramidal tract cell of a monkey is shown during periods when the animal was awake and asleep. It can be seen that the description of such spike trains, short of enumerating all the occurrence times of impulses, is difficult. Usually the collection of interspike intervals is collected into a histogram; examples from another experiment are shown in Figure 9.3.

Given that a sequence of occurrence times of action potentials characterizes the activity of a nerve cell, it is then of interest to see how the train of impulses changes under various conditions. Since the spike trains themselves and the dynamical processes that cause them are usually random, it is clear that models of the activity of real neurons must be *stochastic* and not deterministic. The introduction of randomness makes the mathematical theory more difficult and progress has been made only with what are physically very simple models. Previous works dealing with the stochastic aspects of neural activity are those of Goel and Richter-Dyn (1974), Holden (1976b), MacGregor and Lewis (1977), Ricciardi (1977), and Sampath and Srinivasan (1977).

There have been numerous experimental studies of the stochastic activity of nerve cells. We will mention some representative ones but do not have space to describe the results. In the investigations of the following list, the focus is sometimes on the spontaneous activity or

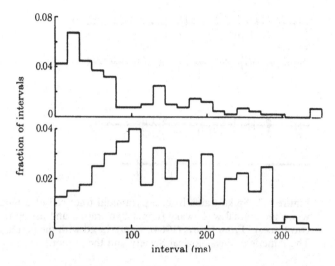

Figure 9.3. Histograms of ISIs for a cell in the cat visual cortex when the animal was "sitting in complete darkness awake, but relaxed" (upper figure) and when it was "alarmed by a hissing noise." [From Burns and Webb (1976). Reproduced with the permission of The Royal Society and the authors.]

on the effects of various influences such as state of arousal, learning, drugs, sensory inputs, epileptogenesis, radiation, and so forth. For studies on *muscle spindles*, in addition to the references already given, see Stein and Matthews (1965); on cells of the *auditory system*, see Gerstein and Kiang (1960), Grossman and Viernstein (1961), Rodieck, Kiang, and Gerstein (1962), Goldberg, Adrian, and Smith (1964), Pfeiffer and Kiang (1965), Gerstein, Butler, and Erulkar (1968), Molnar and Pfeiffer (1968), Kiang and Moxon (1974), Johnson and Kiang (1976), and Ryan and Miller (1977); on cells in the *visual pathway*, see Levick et al. (1961), Bishop, Levick, and Williams (1964), and Burke and Sefton (1966a, b); on cells in the *cerebral cortex* including pyramidal cells, see Evarts (1964), Koike et al. (1970), Whitsel, Roppolo, and Werner (1972), O'Brien, Packham, and Brunnhoelzl (1973), Steriade, Wyzinski, and Apostol (1973), Wyler and Fetz (1974), Burns and Webb (1976), Webb (1976a, b), Whitsel, Schreiner, and Essick (1977), and Schreiner, Essick, and Whitsel (1978); on *hippocampal pyramids*, see Bassant (1976); on *chemoreceptors*, see Silk and Stein (1966); on cells of the *dorsalspinocerebellar tract*, see Jansen, Nicolaysen, and Rudjord (1966) and Pyatigorskii (1966); on *spinal motoneurons*, see Calvin and Stevens (1968); on cells of the *thalamus*, see Poggio and Viernstein (1964), Nakahama et al.

(1966), Baker (1971), Lamarre, Filion, and Cordeau (1971), Dormont (1972), and Benoit and Chataigner (1973); on cells of the *amygdaloid complex*, see Eidelberg, Goldstein, and Deza (1967); on *Purkinje cells of the cerebellum*, see Braitenberg et al. (1965), Pellet et al. (1974), and Woodward, Hoffer, and Altman (1974); on cells in certain *invertebrates*, see Firth (1966), Junge and Moore (1966), Buno, Fuentes, and Segundo (1978), and Holden and Ramadan (1979); on cells of the *vestibular system*, see Wylie and Pelpel (1971), Goldberg, Fernandez, and Smith (1982), and Goldberg, Smith, and Fernandez (1984); and on cells of the *reticular formation*, see Sparks and Travis (1968) and Syka, Popelar, and Radil-Weiss (1977). In addition, in some experiments nerve cells have been deliberately subjected to random stimulation (Wilson and Wyman 1965; Redman and Lampard 1967, 1968; Redman, Lampard, and Annal 1968; Lampard and Redman 1969; Guttman, Feldman, and Lecar 1974; Bryant and Segundo 1976).

Synaptic noise

Noise has also been observed in other nerve-cell preparations. Fatt and Katz (1952) observed randomly occurring synaptic potentials called miniature endplate potentials (m.e.p.p.'s) at the frog neuromuscular function. Their amplitudes were on the order of $0.5\,mV$ and their frequencies were between 0.1 to 100 impulses/s for various muscle fibers. These spontaneous potentials have been important in helping elucidate the nature of synaptic transmission and are studied in Sections 9.3 and 9.4. Spontaneous miniature EPSP's and IPSP's have also been observed in spinal motoneurons of the frog (Katz and Miledi 1963) and the cat (Calvin and Stevens 1968) as well as in pyramidal cells of the cat motor cortex (Watanabe and Creutzfeldt 1966). Sometimes these spontaneous postsynaptic potentials are referred to as *synaptic noise*.

Membrane noise

Even when apparently steady conditions prevail, there are observed small fluctuations in the electrical potential across the nerve-cell membrane. These fluctuations have been attributed to the back-and-forth motion of ions and electrons due to thermal agitation (*Brownian motion*), the discrete nature of currents through the membrane (*shot noise*), and conductance changes induced by the random opening and closing of ion channels. These random fluctuations in membrane potential may be collectively referred to as *membrane noise*.

9.2   Probability and random variables

Probability enters modeling when we do not know, or cannot prescribe with certainty, the conditions or mechanisms that prevail in the system of interest. For example, it is not possible to predict the times of occurrence of the action potentials in Figure 9.2. On the basis of an ISI histogram all we can do, on the basis of collected data, is make statements of the kind, "there is a 40% chance that a spike will occur in the next 10 ms." Statements such as this can be collected into more precise statements in terms of *random variables* and *random processes*. We will give a brief review of some of the basic concepts – texts such as Feller (1968) and Chung (1979) should be consulted for thorough introductory treatments. A wide range of applications is considered in Tuckwell (1988a).

When an experiment is performed whose outcome is uncertain, the collection of possible elementary outcomes is called a *sample space*. Roughly speaking, a *random variable* is an observable that takes on numerical values with certain probabilities. It is a real-valued function defined on the elements of a sample space. We adopt the convention that random variables themselves are denoted by capital letters, whereas the values they take on are denoted by lowercase letters.

Discrete random variables take on finitely many or countably infinitely many values. The most important such random variables for us are binomial and Poisson.

A *binomial* random variable X has the probability law

$$p_k = \Pr\{ X = k \} = \binom{n}{k} p^k q^{n-k}, \qquad k = 0, 1, 2, \ldots, n, \quad (9.3)$$

where $0 \le p \le 1$, $q = 1 - p$, and n is a positive integer. The number of heads in n tosses of a coin that has probability p of landing heads has this probability law.

A *Poisson* random variable with parameter $\lambda > 0$ takes on non-negative integer values and has the probability law

$$p_k = \Pr\{ X = k \} = \frac{e^{-\lambda}\lambda^k}{k!}, \qquad k = 0, 1, 2, \ldots. \quad (9.4)$$

For both of these random variables the total probability mass is unity

$$\sum_k p_k = 1. \quad (9.5)$$

Continuous random variables take on a continuum of values. For any random variable the *distribution function* is

$$F(x) = \Pr\{ X \le x \}. \quad (9.6)$$

Usually the probability law of a continuous random variable can be expressed through its *probability density* $p(x)$, which is the derivative of its distribution function. Then, roughly speaking, $p(x)\,dx$ is the probability that the random variable takes on values in $(x, x + dx]$. In what follows two kinds of continuous random variables are important. One is the *normal*, or *Gaussian*, random variable with density

$$p(x) = \frac{1}{\sqrt{2\pi\sigma^2}}\exp\left[-\frac{(x-\mu)^2}{2\sigma^2}\right], \qquad -\infty < x < \infty, \quad (9.7)$$

where μ and σ^2 are constants. The other has a *gamma* density

$$p(x) = (\lambda/\Gamma(r))(\lambda x)^{r-1}e^{-\lambda x}, \qquad x > 0, \tag{9.8}$$

where $\lambda > 0$ and $r > 0$ are constants and $\Gamma(r) = \int_0^\infty e^{-x}x^{r-1}\,dx$ is the gamma function. A special case is that of an *exponentially distributed* random variable with $r = 1$ so the density is $p(x) = \lambda e^{-\lambda x}$. For continuous random variables the total probability is also unity

$$\int p(x)\,dx = 1, \tag{9.9}$$

where the range of integration is $(-\infty, \infty)$ for a normal random variable and $(0, \infty)$ for a gamma variate.

Mean, variance, and covariance

The *mean*, or *expected value*, of an integer-valued random variable is

$$E[X] = \sum_k kp_k. \tag{9.10}$$

This gives $E[X] = np$ for the binomial and $E[X] = \lambda$ for the Poisson random variable.

For continuous random variables,

$$E[X] = \int xp(x)\,dx. \tag{9.11}$$

This gives $E[X] = \mu$ for the normal and $E[X] = r/\lambda$ for the gamma random variable. Note that expectation is a linear operation, so $E[aX + bY] = aE[X] + bE[Y]$.

The *second moment* is $E[X^2]$ and the *variance*, which tells us how much scatter there is about the mean, is

$$\mathrm{Var}[X] = E\left[(X - E[X])^2\right] = E[X^2] - E^2[X]. \tag{9.12}$$

The variances of the random variables considered above are: binomial, npq; Poisson, λ; normal, σ^2; and gamma, r/λ^2.

Conditional probability and independence

Let A and B be two random events. The *conditional probability* of A given B is defined as

$$\Pr\{A|B\} = \Pr\{AB\}/\Pr\{B\}, \tag{9.13}$$

where AB means both A and B occur and it is assumed that $\Pr\{B\} \neq 0$. That is, only those occurrences of A that are simultaneous with those of B are taken into account. This extends to random variables. For example, if X and Y are two random variables, defined on the same sample space, taking on values x_i, $i = 1,2$, and y_j, $j = 1,2,\ldots$, respectively, then the conditional probability that $X = x_i$ given $Y = y_j$ is

$$\Pr\{X = x_i | Y = y_j\} = \Pr\{X = x_i, Y = y_j\}/\Pr\{Y = y_j\}. \tag{9.14}$$

Note that (9.13) rearranges to $\Pr\{AB\} = \Pr\{B\}\Pr\{A|B\}$.

The *conditional expectation* of X given $Y = y_j$ is

$$E\left[X|Y = y_j\right] = \sum_i x_i \Pr\{X = x_i | Y = y_j\}. \tag{9.14A}$$

The expected value of XY is

$$E[XY] = \sum_{i,j} x_i y_j \Pr\{X = x_i, Y = y_j\}, \tag{9.15}$$

and the *covariance* of two random variables X and Y is

$$\mathrm{Cov}[X,Y] = E\left[(X - E[X])(Y - E[Y])\right] = E[XY] - E[X]E[Y]. \tag{9.16}$$

The covariance is a measure of the linear dependence of X and Y.

If X and Y are *independent*, the value of Y should have no influence on the probability that X takes on its values. Hence we may define X and Y as independent if

$$\Pr\{X = x_i | Y = y_j\} = \Pr\{X = x_i\}, \tag{9.17}$$

for all i, j. Equivalently, $\Pr\{X = x_i, Y = y_j\} = \Pr\{X = x_i\}\Pr\{Y = y_j\}$, which leads, in the case of independent random variables, to

$$E[XY] = E[X]E[Y]. \tag{9.17A}$$

It also follows that if X and Y are independent, $\mathrm{Cov}[X,Y] = 0$ (but note that $\mathrm{Cov}[X,Y] = 0$ does not always imply that X and Y are independent). If X_i, $i = 1,2,\ldots,n$, are mutually independent, then

$$\mathrm{Var}\left[\sum_{i=1}^{n} X_i\right] = \sum_{i=1}^{n} \mathrm{Var}[X_i]. \tag{9.17B}$$

If $A_i, i = 1, 2, \ldots, n$, are mutually exclusive events and at least one must occur, and B is another event, then

$$\Pr\{B\} = \sum_{i=1}^{n} \Pr\{A_i\}\Pr\{B|A_i\}, \qquad (9.17C)$$

which is the *law of total probability*.

Finally, the *characteristic function* of a random variable X is defined as the expectation of the complex random variable $\exp(iuX)$, where u varies from $-\infty$ to $+\infty$:

$$\phi_X(u) = E[e^{iuX}], \qquad u \in (-\infty, \infty). \qquad (9.18)$$

(Here $i = \sqrt{-1}$.) Characteristic functions are important because there is a one-to-one correspondence between them and distribution functions.

It is left as an exercise to show that if X is Poisson with parameter λ, then

$$\phi_X(u) = \exp[\lambda(e^{iu} - 1)], \qquad (9.18A)$$

whereas if X is normal with mean μ and variance σ^2,

$$\phi_X(u) = \exp[i\mu u - \tfrac{1}{2}u^2\sigma^2]. \qquad (9.18B)$$

9.3   The quantum hypothesis in synaptic transmission

In this section we will obtain a probabilistic description of the small voltage changes that occur at spontaneously active synapses. This will be done in the context of nerve–muscle synapses but the same kind of stochastic model should apply at synapses within the nervous system.

In Section 9.1 we mentioned Fatt and Katz's (1952) discovery of randomly occurring miniature endplate potentials at the frog neuromuscular junction. An example of the records they obtained is shown in Figure 9.4. The mean amplitude of the m.e.p.p.'s was about 0.5 mV. This should be compared with the normal postsynaptic response, the endplate potential (e.p.p.), which results when a nerve impulse invades the presynaptic terminal. The e.p.p. has an amplitude between 50 and 70 mV (Kuffler and Nicholls 1976).

In a reduced Ca^{2+} bathing solution the amplitude of the e.p.p. is reduced. Fatt and Katz observed that in such low Ca^{2+} solutions the amplitudes of the reduced e.p.p.'s were approximately in multiples of the amplitudes of the spontaneous m.e.p.p.'s. The *quantum hypothesis* (del Castillo and Katz 1954) was advanced that the normal e.p.p. is the result of the almost simultaneous occurrence of several m.e.p.p.'s.

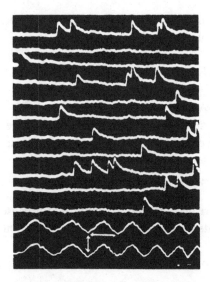

Figure 9.4. Randomly occurring postsynaptic potentials (m.e.p.p.'s) recorded intracellularly from frog muscle fiber. Several records are shown from the same fiber. [From Fatt and Katz (1952). Reproduced with the permission of The Physiological Society and the authors.]

It was speculated that the transmitter (acetylcholine) was, in fact, released in packets containing several thousand molecules corresponding to the quantal EPSP's mentioned in Section 1.5. Furthermore, it seemed likely that the *synaptic vesicles* of diameter about 500 Å, seen with the electron microscope to reside in the presynaptic terminal, were, in fact, the packets of transmitter.

Probability model

We assume there are n sites within the nerve terminal at which transmitter release may occur. When an action potential invades the terminal, release occurs at a random number M of sites. The amplitude of the response due to release at each site is random, that at the ith active site being V_i. Each of the V_i's is assumed to have the same probability distribution and the sites act independently of each other. The amplitude of the e.p.p. is then

$$V = V_1 + V_2 + \cdots + V_M. \tag{9.19}$$

This is a sum in which the number of terms is random.

In the first instance we would assume that M is binomial with parameters n and p, where p is the probability that a site is active. However, if p is small we may use the *Poisson approximation* to the

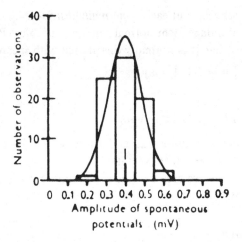

Figure 9.5. Histogram of the amplitudes of the spontaneous
m.e.p.p.'s. The smooth curve is a normal density. [From Boyd and
Martin (1956). Reproduced with the permission of The Physiological
Society and the authors.]

binomial, which has the advantage that now the distribution of M
contains only one parameter, $\lambda = np$, the mean number of active sites.
The value of λ can be estimated from the fraction of trials, which
result in no response, that is,

$$\Pr\{M = 0\} = e^{-\lambda}. \tag{9.20}$$

When M has a Poisson distribution, the random sum V is said to
have a *compound Poisson distribution*.

Figure 9.5 shows the histogram of amplitudes of the miniature
spontaneous potentials in one preparation. The smooth curve through
the histogram is a normal probability density with the same mean and
variance. Thus we assume that the V_i's are normal with mean μ and
variance σ^2. We are now able to find the density of the amplitude of
the e.p.p. V.

The *law of total probability* gives

$$\Pr\{v < V < v + dv\} = \sum_{m=0}^{\infty} \Pr\{v < V < v + dv \mid M = m\}\Pr\{M = m\} \tag{9.21}$$

The conditional probability that $V \in (v, v + dv)$ given $M = m$ is found
as follows. If $M = m > 0$, then there are exactly m terms in the sum
(9.19)

$$V = V_1 + V_2 + \cdots + V_m. \tag{9.22}$$

These terms are *independent* and each is *normally distributed*. A basic theorem on the sum of independent normal random variables (Parzen 1962, page 17) tells us that V is normally distributed with mean

$$E[V|M = m] = mE[V_1] = m\mu, \tag{9.23}$$

and variance

$$\text{Var}[V|M = m] = m\,\text{Var}[V_1] = m\sigma^2. \tag{9.24}$$

In the case $m = 0$ the amplitude has a density, which is a delta function concentrated at $v = 0$ with weight $e^{-\lambda}$.

Putting all this together, we arrive at the probability density of the amplitude of the e.p.p.,

$$p_V(v) = \exp(-\lambda)\left\{ \delta(v) + \frac{1}{\sqrt{2\pi\sigma^2}} \sum_{m=1}^{\infty} \frac{\lambda^m}{m!\sqrt{m}} \exp\left[\frac{-(v - m\mu)^2}{2m\sigma^2} \right] \right\}, \tag{9.25}$$

which agrees with the formula of Bennett and Florin (1974). It is left as an exercise to show that

$$E[V] = \lambda\mu, \tag{9.25A}$$

$$\text{Var}[V] = \lambda(\mu^2 + \sigma^2). \tag{9.25B}$$

Figure 9.6 shows the histogram of amplitudes of the endplate potentials along with the density predicted by (9.25). It can be seen that there is good agreement between the experimental and theoretical result, thus substantiating the quantum hypothesis. Note that since the normal e.p.p. is about 50 mV, then as many as 100 or more quanta are released when the action potential invades the terminal. Supporting evidence for the quantum hypothesis has been obtained in other preparations [see Martin (1977)]. However, a gamma, rather than a normal, density has sometimes been needed for the unit response [see, for example, Bornstein (1978)] and a binomial analysis has sometimes been required when the Poisson approximation has proven inadequate (Miyamoto 1975; Volle and Branisteanu 1976). Brown, Perkel, and Feldman (1976) showed, with the aid of computer simulation, that spatial and temporal nonuniformities in the release parameters can easily lead to erroneous estimates of these quantities.

9.4   The Poisson process

A *random process* is a family of random variables. We will only be concerned with families of random variables parameterized by a continuous index t, the time. The chief physical random processes

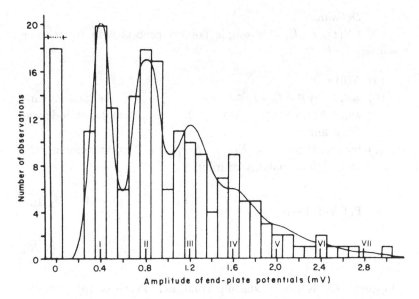

Figure 9.6. Histogram of amplitudes of the e.p.p. The smooth curve is the density of a compound Poisson random variable as given by (9.25). [From Boyd and Martin 1956). Reproduced with the permission of The Physiological Society and the authors.]

with which we will be concerned are the nerve membrane potential and the number of synaptic inputs.

A random variable has a fixed probability law, which is easy to describe, but the probability laws of continuous-time random processes are complicated. If we let $X(t)$ denote the value of a general random process at time t, the whole process is the family $\{ X(t), t \geq 0 \}$, which we sometimes abbreviate to X.

The first random process we examine is the *simple Poisson process*. This is important for several reasons:

(i) it is a basic, much studied process and can be used as a standard against which some physical processes can be compared;

(ii) it may be used to synthesize more complicated processes of interest;

(iii) it forms a useful approximation for the inputs to a cell when these are many and unsynchronized.

There are many equivalent definitions of the Poisson process [see, for example, Parzen (1962)] one of which is as follows.

Definition

$\{N(t),\ t \leq 0\}$ is a simple Poisson process with *intensity* or *mean rate* λ if:

(a) $N(0) = 0$;

(b) given any $0 = t_0 < t_1 < t_2 < \cdots < t_{n-1} < t_n$, the random variables $N(t_k) - N(t_{k-1})$, $k = 1, 2, \ldots, n$, are mutually independent; and

(c) for any $0 \leq t_1 < t_2$, $N(t_2) - N(t_1)$ is a Poisson random variable with probability distribution

$$\Pr\{N(t_2) - N(t_1) = k\} = \frac{(\lambda(t_2 - t_1))^k \exp(-\lambda(t_2 - t_1))}{k!},$$

$$k = 0, 1, 2, \ldots. \quad (9.26)$$

Property (a) is just a starting condition. Property (b) puts the Poisson process in the class of processes with *independent increments*. Property (c) tells us that the increments are *stationary* (since only time differences matter) with Poisson distributions. The meaning of λ will become clear shortly.

From (9.26) with $t_1 = 0$ and $t_2 = t$, we see that $N(t)$ is a Poisson random variable with mean and variance equal to λt. Also, with $t_1 = t$ and $t_2 = t + \Delta t$, we find

$$\Pr\{N(t + \Delta t) - N(t) = k\} = \frac{(\lambda \Delta t)^k \exp(-\lambda \Delta t)}{k!}$$

$$= \begin{cases} 1 - \lambda \Delta t + o(\Delta t), & k = 0, \\ \lambda \Delta t + o(\Delta t), & k = 1, \\ o(\Delta t), & k \geq 2, \end{cases}$$

$$(9.27)$$

where $o(\Delta t)$ means terms that, as $\Delta t \to 0$, approach zero faster than Δt itself. Hence, in very small time intervals, the process is most likely to stay unchanged ($k = 0$) or undergo a step increase of unity ($k = 1$). The value of $N(t)$ will be the number of unit step changes that have occurred in $(0, t]$. A typical *realization* (*sample path*, *trajectory*) of the process will appear as sketched in Figure 9.7A.

In Figure 9.7B we have inserted a cross on the t-axis at the times when $N(t)$ jumps by unity. Each cross may be associated with the

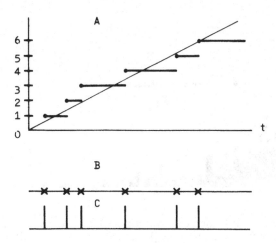

Figure 9.7. A – Realization of a Poisson process. B – Associated realization of the Poisson point process. C – Hypothetical spike train.

occurrence of a certain kind of event such as a postsynaptic potential or an action potential. The crosses (points) form a realization of a *Poisson point process* and $N(t)$ thus records or counts the number of events in $(0, t]$. Figure 9.7C shows a hypothetical spike train, which can be associated with the point process.

The waiting time to an event

Consider any $s > 0$ and let T_1 be the time to the first event occurring after s. Then we find that T_1 *is exponentially distributed with mean* $1/\lambda$.

Proof. The probability that one has to wait longer than t for the first event is the probability that there are no events in $(s, s + t]$. Thus

$$\Pr\{T_1 > t\} = \Pr\{N(s+t) - N(s) = 0\} = e^{-\lambda t}, \qquad t > 0. \tag{9.28}$$

Thus the distribution function of T_1 is $1 - e^{-\lambda t}$ and hence the probability density function p_1 of T_1 is

$$p_1(t) = \lambda e^{-\lambda t}, \qquad t > 0, \tag{9.29}$$

as required. Since s was completely arbitrary, it could have coincided with the time of an event. It may be shown, in fact, that *the time interval between events is exponentially distributed with mean* $1/\lambda$. Since the average waiting time between events is $1/\lambda$, there are, roughly

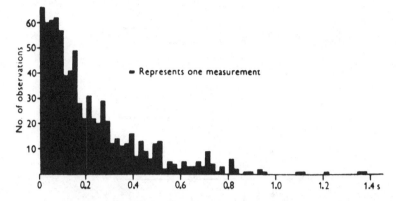

Figure 9.8. Histogram of time intervals between m.e.p.p.'s at the frog neuromuscular junction. [From Fatt and Katz (1952). Reproduced with the permission of The Physiological Society and the authors.]

speaking, on average λ events per unit time. Thus λ is called the mean rate or intensity.

Figure 9.8 shows the histogram, obtained by Fatt and Katz (1952), of time intervals between the spontaneous miniature endplate potentials at the frog neuromuscular junction. The histogram has the shape of an exponential distribution. Fatt and Katz were thus led to make their *Poisson hypothesis*, that the arrival times of the m.e.p.p.'s constituted a Poisson point process [see Van de Kloot, Kita, and Cohen (1975)].

The Poisson process as a primitive model for nerve-cell activity

Let the depolarization of a nerve cell be $\{V(t),\ t \geq 0\}$. Suppose that excitatory inputs occur at random in accordance with events in a simple Poisson process $\{N(t),\ t \geq 0\}$ with mean rate λ. Each excitatory input causes V to increase by a_E. When V reaches or exceeds the constant threshold level $\theta > 0$, the cell emits an action potential. Then

$$V(t) = a_E N(t), \qquad V < \theta, V(0) = 0. \tag{9.30}$$

In this primitive nerve-cell model, what is the probability distribution of the time interval between action potentials?

To answer this we first ask what is the waiting time T_k until the kth event in a simple Poisson process after the arbitrary time point s. We will show that T_k *has a gamma density with parameters k and λ*.

Proof. The kth event will occur in $(s+t, s+t+\Delta t]$ if and only if there are $k-1$ events in $(s, s+t]$ and one event in $(s+t, s+t+\Delta t]$.

It follows that

$$\Pr\{T_k \in (t, t+\Delta t]\} = \frac{e^{-\lambda t}(\lambda t)^{k-1}\lambda \Delta t}{(k-1)!} + o(\Delta t), \qquad k = 1, 2, \ldots .$$

$$(9.31)$$

Hence the density of T_k is

$$p_k(t) = \frac{\lambda(\lambda t)^{k-1} e^{-\lambda t}}{(k-1)!}, \qquad t > 0, \qquad (9.32)$$

as required.

Hence we find that *the waiting time for the kth event has a gamma density with parameters k and λ.* Thus T_k has mean k/λ and variance k/λ^2. Some gamma densities are illustrated in Figure 9.9.

To return to the primitive nerve-cell model, an action potential is emitted when V reaches or exceeds θ, or, equivalently, when N reaches or exceeds θ/a_E. Letting $[x]$ denote the largest integer less than x we find that $1 + [\theta/a_E]$ excitatory inputs are required. *Hence the time interval between action potentials has a gamma density with parameters $1 + [\theta/a_E]$ and λ.* The mean time interval between action potentials is $(1 + [\theta/a_E])/\lambda$.

The gamma densities, in fact, resemble the ISI histograms obtained for many nerve cells and have often been fitted to them [see, for example, Stein (1965)]. However, the model employed to derive the gamma densities incorporates no decay of membrane potential between excitatory inputs. Only when the cell has an *exceedingly large*

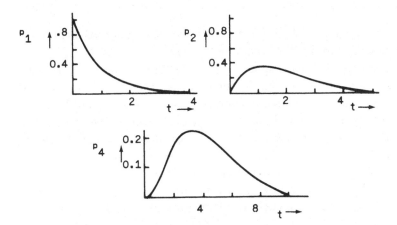

Figure 9.9. Gamma densities for $\lambda = 1$ and $k = 1, 2,$ and 4. [Adapted from Cox and Miller (1965).]

time constant and the rate of incoming excitation is very fast will this approximation be valid. Finally, we note that when θ/a_E is large the gamma density becomes approximately that of a normal random variable.

9.5 Poisson excitation with Poisson inhibition

We will here consider another primitive model for nerve-cell activity in which excitation and inhibition arrive according to independent Poisson processes. This gives what is commonly called a *birth-and-death process*, which in this case Feller calls a *randomized random walk*. Much of what follows is taken from Feller's treatment (1966). Again we ignore the decay of membrane potential between inputs. Despite the unphysiological nature of the model, it is useful because:

(i) the model can be analyzed completely thereby providing a standard with which to compare other models and also real nerve cells; and

(ii) in a limiting case we obtain a Wiener process (see the next section), which is useful in many situations.

Let $V(t)$, $t \geq 0$, be the depolarization at time t. We assume that the number of excitatory inputs in $(0, t]$ is $N_E(t)$, where N_E is a Poisson process with mean rate λ_E, and that the number of inhibitory inputs is $N_I(t)$, where N_I is a Poisson process with mean rate λ_I. Each excitatory input makes V jump up by unity, whereas each inhibitory input causes V to jump down by unity. Thus

$$V(t) = N_E(t) - N_I(t), \qquad V(0) = 0, V < \theta. \tag{9.33}$$

The first thing we find is the probability distribution of $V(t)$ when there is no threshold for action potentials.

Consider what may happen in $(t, t + \Delta t]$. A jump in N_E occurs with probability $\lambda_E \Delta t + o(\Delta t)$, a jump in N_I occurs with probability $\lambda_I \Delta t + o(\Delta t)$, and with probabilities $1 - \lambda_E \Delta t + o(\Delta t)$ and $1 - \lambda_I \Delta t + o(\Delta t)$, respectively, N_E and N_I remain unchanged. The probability of a jump of either kind in V in $(t, t + \Delta t]$ is thus $(\lambda_E + \lambda_I)\Delta t + o(\Delta t)$ and the probability of no jump is $(1 - \lambda_E \Delta t + o(\Delta t))(1 - \lambda_I \Delta t + o(\Delta t)) = 1 - (\lambda_E + \lambda_I)\Delta t + o(\Delta t)$. The times at which V changes are thus a Poisson point process with mean rate $\lambda = \lambda_E + \lambda_I$. In fact, if we let $N(t)$ be the number of jumps (of either kind) of V in $(0, t]$, then

$$N(t) = N_E(t) + N_I(t), \tag{9.34}$$

and $\{N(t),\ t \geq 0\}$ is a Poisson process with mean rate λ. From the definition of conditional probability we find

$$\Pr(V \text{ jumps by } +1 \text{ in } (t, t + \Delta t] \mid \text{a jump in } V \text{ occurs in } (t, t + \Delta t])$$

$$= \lambda_E / \lambda \doteq p, \qquad (9.35)$$

$$\Pr(V \text{ jumps by } -1 \text{ in } (t, t + \Delta t] \mid \text{a jump in } V \text{ occurs in } (t, t + \Delta])$$

$$= \lambda_I / \lambda \doteq q, \qquad (9.36)$$

and $p + q = 1$.

We seek

$$p_m(t) = \Pr\{V(t) = m \mid V(0) = 0\}, \qquad m = 0, \pm 1, \pm 2, \ldots. \tag{9.37}$$

which is the conditional probability that $V(t) = m$ for an initial value zero. Such a quantity is an example of a *transition probability*, which we associate with a class of processes called *Markov* to which the process V belongs. We will for convenience drop the reference to the initial state and consider $m > 0$.

Let the process V be at m at t and suppose that $n \geq m$ jumps have occurred, of which n_1 were $+1$ and n_2 were -1. Then we must have

$$n = n_1 + n_2, \tag{9.38A}$$

$$m = n_1 - n_2, \tag{9.38B}$$

and hence

$$n_1 = (m + n)/2, \tag{9.38C}$$

$$n = m + 2n_2. \tag{9.38D}$$

The probability that $V(t) = m$ if n jumps have occurred in $(0, t]$ is the probability that a binomial random variable with parameters n and p takes the value n_1. That is,

$$\Pr\{V(t) = m \mid n \text{ jumps in } (0, t]\} = \binom{n}{n_1} p^{n_1} q^{n - n_1}$$

$$= \binom{n}{\dfrac{n + m}{2}} p^{(n+m)/2} q^{(n-m)/2}. \tag{9.39}$$

By the law of total probability,

$$\Pr\{V(t) = m\} = \sum_{n \geq m} \Pr\{V(t) = m \mid n \text{ jumps in } (0, t]\}$$

$$\times \Pr\{n \text{ jumps in } (0, t]\}. \tag{9.40}$$

Since n jumps in $(0, t]$ has probability $e^{-\lambda t}(\lambda t)^n/n!$, we find

$$p_m(t) = e^{-\lambda t} \sum_{n=m}^{\infty}{}' \frac{(\lambda t)^n}{n!} \binom{n}{\frac{n+m}{2}} p^{(n+m)/2}q^{(n-m)/2}, \quad (9.41)$$

where the prime on the summation sign indicates that summation is over either even or odd n depending on whether m is even or odd, respectively.

Utilizing (9.38) and the fact that $n = m, m+2, m+4, \ldots$ implies $n_2 = 0, 1, 2, \ldots$, this becomes

$$p_m(t) = e^{-\lambda t} \sum_{n_2=0}^{\infty} \frac{(\lambda t)^{m+2n_2}}{(m+2n_2)!} \binom{m+2n_2}{m+n_2} p^{m+n_2}q^{n_2}. \quad (9.42)$$

In terms of the modified Bessel function,

$$I_\rho(x) = \sum_{k=0}^{\infty} \frac{1}{k!\,\Gamma(k+\rho+1)} \left(\frac{x}{2}\right)^{2k+\rho}, \quad (9.43)$$

we get

$$p_m(t) = \left(\frac{\lambda_E}{\lambda_I}\right)^{m/2} e^{-\lambda t} I_m\!\left(2t\sqrt{\lambda_E\lambda_I}\right).$$

9.5.1 Time of first passage to threshold

We assume that there is a fixed threshold θ, which when reached by V leads to the emission of an action potential. The threshold condition is an imposed one and after a spike the potential is artificially reset to zero, possibly after a dead time or refractory period. Passage to time-varying thresholds in this model does not seem to have been considered.

We let θ be a positive integer and seek the *time of first passage of V to θ*, which is identified with the interspike interval. To find the probability distribution of the first-passage time, we employ the *method of images* in the symmetric case ($\lambda_E = \lambda_I$) and the *renewal equation* in the asymmetric case.

(A) *Symmetric case: method of images*

We will first find $p_m^*(t)$, the probability that the randomized random walk is at level m at t but has stayed below θ up to time t. This is in distinction to $p_m(t)$, which includes passages below, to, and above θ. Consider Figure 9.10, where a randomized walk process U is shown starting from the image point 2θ and having the value m at t. There is a one-to-one correspondence between such paths of U and

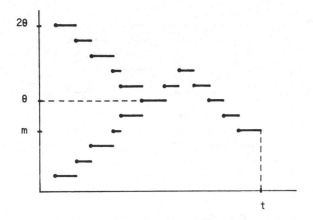

Figure 9.10. Paths considered in the method of images.

those of V that start at 0, touch and/or cross the level θ in $(0, t)$, and end up at m at t. By symmetry the probability assigned to such paths is $p_{2\theta - m}(t)$. These paths are excluded in computing p_m^* so we have

$$p_m^*(t) = p_m(t) - p_{2\theta - m}(t). \tag{9.44}$$

To obtain the probability density $f_\theta(t)$ of the time of first passage to level θ, note that a first passage to θ occurs in $(t, t + \Delta t]$ if V has stayed below θ in $(0, t]$, is, in fact, at $\theta - 1$ at t, and a jump of $+1$ occurs in $(t, t + \Delta t]$. The probability of a jump of $+1$ in $(t, t + \Delta t]$ is $\lambda_E \Delta t = (\lambda/2)\Delta t$. Putting these probabilities together gives

$$f_\theta(t)\Delta t = p_{\theta - 1}^*(t)(\lambda/2)\Delta t. \tag{9.45}$$

Utilizing (9.43) and (9.44) and the fact that $\lambda_E = \lambda_I$, we find

$$f_\theta(t) = \frac{\lambda}{2} e^{-\lambda t}[I_{\theta - 1}(2\lambda_E t) - I_{\theta + 1}(2\lambda_E t)], \qquad t > 0. \tag{9.46}$$

A more succinct expression results on using the recurrence relation for the modified Bessel function,

$$I_{\theta - 1}(x) - I_{\theta + 1}(x) = (2\theta/x)I_\theta(x), \tag{9.47}$$

whereupon

$$f_\theta(t) = (\theta/t)e^{-\lambda t}I_\theta(2\lambda_E t), \qquad t > 0. \tag{9.48}$$

(B) *General case: the renewal equation*

When the rates of arrival of jumps up and down are not equal, we resort to another method for obtaining $f_\theta(t)$. The idea on which the method is based is illustrated in Figure 9.11. A path is shown starting at zero and attaining the value $m > \theta$ at t. Since $m > \theta$,

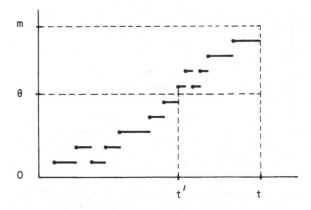

Figure 9.11. Paths that lead to the renewal equation.

such a path must have at some time before t passed through the level θ and, in particular, at some time $t' < t$ done this for the first time. Integrating over all such paths gives, using a continuous version of the law of total probability,

$$p_m(t) = \int_0^t f_\theta(t') p_{m-\theta}(t-t') \, dt'. \qquad (9.49)$$

This integral equation is called a *renewal equation*, which we will solve using Laplace transform methods. Note that the probability of a transition from θ at t' to m at t is the same as the probability of a transition from 0 at time zero to $m - \theta$ at $t - t'$ (spatial and temporal homogeneity).

Equation (9.49) is a *convolution* of f_θ and $p_{m-\theta}$. From Table 3.2 we see that the Laplace transform of the convolution of two functions is the product of their Laplace transforms. Thus, denoting Laplace transforms by the extra subscript L,

$$p_{m,L}(s) = f_{\theta,L}(s) p_{m-\theta,L}(s), \qquad m < \theta, \qquad (9.50)$$

where s is the transform variable. Rearranging this gives the following useful relation:

$$f_{\theta,L}(s) = p_{m,L}(s)/p_{m-\theta,L}(s). \qquad (9.51)$$

The transforms on the right can be found as series. From (9.42) and Table 3.2, we find

$$\mathcal{L}\{e^{\lambda t} p_m(t)\} = \mathcal{L}\left\{ \sum_{n_2=0}^{\infty} \frac{(\lambda t)^{m+2n_2}}{(m+2n_2)!} \binom{m+2n_2}{m+n_2} p^{m+n_2} q^{n_2} \right\}$$

$$= \frac{\lambda^m p^m}{s^{m+1}} \sum_{n_2=0}^{\infty} \left(\frac{\lambda pq}{s^2} \right)^{n_2} \binom{m+2n_2}{m+n_2}. \qquad (9.52)$$

Utilizing the property $\mathscr{L}\{e^{ct}f(t)\} = f_L(s-c)$, we find

$$f_{\theta,L}(s) = \frac{\lambda^\theta p^\theta}{(s+\lambda)^\theta} \frac{\sum\limits_{n_2=0}^{\infty} \left(\dfrac{\lambda pq}{(s+\lambda)^2}\right)^{n_2} \binom{m+2n_2}{m+n_2}}{\sum\limits_{n_2=0}^{\infty} \left(\dfrac{\lambda pq}{(s+\lambda)^2}\right)^{n_2} \binom{m-\theta+2n_2}{m-\theta+n_2}}.$$

$$(9.53)$$

It is left as an exercise to show that this is the Laplace transform of the required first-passage time density,

$$f_\theta(t) = \theta \left(\frac{\lambda_E}{\lambda_I}\right)^{\theta/2} \frac{e^{-(\lambda_E+\lambda_I)t}}{t} I_\theta\left(2t\sqrt{\lambda_E\lambda_I}\right), \qquad t>0. \quad (9.54)$$

Moments of the firing time

Let T_θ be the (random) time taken for V to reach θ from the initial state zero (resting state). Then T_θ has the probability density function f_θ. The nth moment of T_θ is

$$\mu_n = \int_0^\infty t^n f_\theta(t)\, dt. \tag{9.55}$$

When $n=0$ we obtain the total probability mass of T_θ concentrated on $(0,\infty)$. That is,

$$\mu_0 = \Pr\{T_\theta < \infty\} = \int_0^\infty f_\theta(t)\, dt. \tag{9.56}$$

A series representation of this probability can be found but it is difficult to sum the series. However, by applying Theorem 7.1 of Karlin and Taylor (1975) on the probability of extinction in a general birth-and-death process, we find

$$\Pr\{T_\theta < \infty\} = \begin{cases} 1, & \lambda_E \geq \lambda_I, \\ \left(\dfrac{\lambda_E}{\lambda_I}\right)^\theta, & \lambda_E < \lambda_I \end{cases}. \tag{9.57}$$

Thus, if the mean rate of excitation is greater than or equal to the mean rate of inhibition, the time to reach threshold is finite with probability one. On the other hand, if the mean rate of inhibition is greater than that of excitation, the threshold may never be reached, which implies that the neuron may never fire an action potential. Note, however, that this result is obtained in a model that neglects the decay of potential between inputs.

If $\lambda_E < \lambda_I$ the mean firing time is infinite as T_θ has some probability mass at $t=\infty$. When $\lambda_E > \lambda_I$ the mean and variance of the firing time can be found with the aid of the following relation (Gradshteyn and Ryzhik 1965, page 708):

$$\int_0^\infty e^{-\alpha x} I_\nu(\beta x)\, dx = \frac{\beta^\nu}{\sqrt{\alpha^2 - \beta^2}\left(\alpha + \sqrt{\alpha^2 - \beta^2}\right)^\nu}. \tag{9.58}$$

This yields

$$E[T_\theta] = \frac{\theta}{\lambda_E - \lambda_I}, \qquad \lambda_E > \lambda_I, \tag{9.59}$$

$$\mathrm{Var}[T_\theta] = \frac{\theta(\lambda_E + \lambda_I)}{(\lambda_E - \lambda_I)^3}, \quad \lambda_E > \lambda_I. \tag{9.60}$$

The *coefficient of variation*, the standard deviation divided by the mean, is

$$\mathrm{CV}[T_\theta] = \left(\frac{\lambda_E + \lambda_I}{\theta(\lambda_E - \lambda_I)}\right)^{1/2}, \tag{9.61}$$

which indicates that the *coefficient of variation is inversely proportional to the square root of the threshold* for fixed rates of excitation and inhibition. When $\lambda_E = \lambda_I$, although $T_\theta < \infty$ with probability one, the mean (and higher-order moments) of T_θ is infinite.

Note that we have assumed that the excitatory and inhibitory jumps of V are of unit size. If instead the jumps have magnitude a, so that

$$V(t) = a[N_E(t) - N_I(t)], \tag{9.62}$$

then with a threshold $\theta > 0$, not necessarily an integer, the time to get to threshold will be the time for the process with unit jumps to reach $[1 + \theta/a]$.

Tails of the firing-time density

Using the following asymptotic relation for the modified Bessel function at large arguments (Abramowitz and Stegun 1965, page 377),

$$I_\nu(x) \underset{x\to\infty}{\sim} \frac{e^x}{\sqrt{2\pi x}}\left\{1 - \frac{(4\nu^2 - 1)}{8x} + o\left(\frac{1}{x}\right)\right\}, \tag{9.63}$$

we deduce that when there is Poisson excitation and inhibition,

$$f_\theta(t) \underset{t\to\infty}{\sim} \frac{\theta}{2}\left(\frac{\lambda_E}{\lambda_I}\right)^{\theta/2} \frac{1}{\sqrt{\pi(\lambda_E \lambda_I)^{1/2}}} \frac{e^{-t(\sqrt{\lambda_E} - \sqrt{\lambda_I})^2}}{t^{3/2}}$$

$$\times \left\{1 - \frac{4\theta^2 - 1}{16t\sqrt{\lambda_E \lambda_I}} + o\left(\frac{1}{t}\right)\right\}, \tag{9.64}$$

whereas when there is Poisson excitation only, we have the exact result

$$f_\theta(t) = \left[\lambda^\theta/(\theta-1)!\right]t^{\theta-1}e^{-\lambda t}. \tag{9.65}$$

Thus the density of the first-passage time to level θ is quite different in its tails, depending on the presence or absence of inhibition.

9.6 The Wiener process

We will soon proceed to more realistic models, which incorporate the decay of membrane potential between synaptic inputs. Before doing so, we consider an approximation to the process V defined in the previous section. The approximating process is a *Wiener process (or Brownian motion)*, which belongs to the general class of Markov processes called *diffusion processes*. These general concepts will be explained later. Gerstein and Mandelbrot (1964) pioneered the use of the Wiener process in neural modeling.

Diffusion processes have trajectories that are continuous functions of t, in distinction to the randomized random walk whose sample paths are discontinuous. The study of diffusion processes is often less difficult than that of their discontinuous counterparts chiefly because the equations describing their important properties are differential equations about which more is known than differential-difference equations, which arise for discontinuous processes. Among the reasons for studying the Wiener process as a model for nerve membrane potential are:

(i) it is a thoroughly studied process and many of the relevant mathematical problems have been solved; and

(ii) from the Wiener process we may construct many other more realistic models of nerve-cell activity.

The Wiener process as a limiting case of a random walk
Consider the process defined by

$$V_a(t) = a\left[N_E(t) - N_I(t)\right], \qquad t \geq 0, \tag{9.66}$$

where a is a constant, and N_E and N_I are independent Poisson processes with mean rates $\lambda_E = \lambda_I = \lambda$. The process V_a has jumps up and down of magnitude a. We note that

$$E\left[V_a(t)\right] = a\left[\lambda_E - \lambda_I\right]t = 0, \tag{9.67}$$

$$\mathrm{Var}\left[V_a(t)\right] = a^2\left[\mathrm{Var}\left[N_E(t)\right] + \mathrm{Var}\left[N_I(t)\right]\right] = 2a^2\lambda t. \tag{9.68}$$

The characteristic function of $V_a(t)$ is

$$\phi_a(u; t) = E\left[\exp(iuV_a(t))\right]$$

$$= E\left[\exp\left(iua\left(N_E(t) - N_I(t)\right)\right)\right]$$

$$= E\left[\exp\left(iuaN_E(t)\right)\right] E\left[\exp\left(-iuaN_I(t)\right)\right] \qquad (9.69)$$

by the independence of N_E and N_I. From Section 9.2 we find

$$\phi_a(u; t) = \exp\left\{\lambda t\left(e^{iua} + e^{-iua} - 2\right)\right\}. \qquad (9.70)$$

To standardize the random variables $V_a(t)$, we let

$$\lambda = 1/2a^2, \qquad (9.71)$$

so that $V_a(t)$ has mean zero and variance t for all a. It is left as an exercise to show that

$$\lim_{a \to 0} \phi_a(u; t) = e^{-\frac{1}{2}u^2 t} \doteq \phi(u; t). \qquad (9.72)$$

Thus, from Section 9.2, we see that $\phi(u; t)$ is the characteristic function of a normal random variable with mean zero and variance t.

One way to characterize the distance between two random variables is by the differences between their distribution functions. Let $\{X_n, n = 1, 2, \ldots\}$ be a sequence of random variables with distribution function F. If

$$\lim_{n \to \infty} F_n(x) = F(x),$$

for all points x at which F is continuous, we say the sequence $\{X_n\}$ *converges in distribution* to X. We write

$$X_n \overset{d}{\to} X.$$

A basic theorem of probability theory [see, for example, Ash (1970), page 171] tells us that to establish convergence in distribution it is sufficient to prove convergence of the corresponding sequence of characteristic functions. Hence as $a \to 0$, the sequence of random variables $V_a(t)$ converges in distribution to a normal random variable with mean zero and variance t. We let this limiting variable be $W(t)$,

$$V_a(t) \underset{a \to 0}{\longrightarrow} W(t). \qquad (9.73)$$

The process $\{W(t), t \geq 0\}$ is called a *standard Wiener process* and more will be said about the convergence of V_a to W in Section 9.9.

The "standard" refers to the values of the mean and variance,

$$E[W(t)] = 0, \qquad (9.74)$$

$$\mathrm{Var}[W(t)] = t. \qquad (9.75)$$

Definition and some properties of W

In the above limiting procedure, we obtained W by letting the jump size in the random walk become smaller, whereas the rates at which the jumps arrived became faster. Thus W has sample paths, which are, in fact, continuous. So far we have only considered $W(t)$, the value of W at time t. A definition of the process $\{W(t), \, t \geq 0\}$ is the following.

Definition

$\{W(t), \, t \geq 0\}$ is a standard Wiener process (Brownian motion) if:

(a) $W(0) = 0$;
(b) given any $0 \leq t_0 < t_1 < t_2 < \cdots < t_{n-1} < t_n$, the random variables $W(t) - W(t_{k-1})$, $k = 1, 2, \ldots, n$, are independent; and
(c) for any $0 \leq t_1 < t_2$, $W(t_2) - W(t_1)$ is a normal random variable with mean zero and variance $t_2 - t_1$.

Note that $\{V_a(t)\}$ satisfies (a), (b), and (c) asymptotically as $a \to 0$. Thus W shares with the Poisson process the property of having stationary independent increments. The density of $W(t)$ is

$$f_W(x,t) = \frac{1}{\sqrt{2\pi t}} \exp\left\{-\frac{x^2}{2t}\right\}, \qquad -\infty < x < \infty, \, t > 0. \qquad (9.76)$$

An attempt is made in Figure 9.12 to depict some sample paths for W. Although the paths of W are smooth enough to be continuous, they

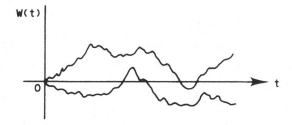

Figure 9.12. A sketch of two sample paths of W.

are, with probability one, nondifferentiable. Nevertheless, the "derivative" of W, denoted by w, is called *white noise* and is a useful concept. Whenever w appears in an equation, an integration is implied.

Wiener process with drift

We may construct new processes from W by multiplying it by a constant and adding a linear drift. Thus

$$X(t) = x_0 + \sigma W(t) + \mu t, \qquad t > 0, \tag{9.77}$$

where $X(0) = x_0$ defines a *Wiener process with variance parameter σ and drift parameter μ*. Linear operations on Gaussian (normal) processes, such as W, produce Gaussian processes. Since

$$E[X(t)] = x_0 + \mu t, \tag{9.78}$$

$$\mathrm{Var}[X(t)] = \sigma^2 t, \tag{9.79}$$

the density of $X(t)$ is

$$f_X(x,t) = \frac{1}{\sqrt{2\pi\sigma^2 t}} \exp\left\{-\frac{(x - x_0 - \mu t)^2}{2\sigma^2 t}\right\}, \qquad -\infty < x < \infty, t > 0. \tag{9.80}$$

As $t \to 0^+$, $f_X(x,t)$ approaches $\delta(x - x_0)$.

The nerve-cell model

It is the Wiener process with drift that Gerstein and Mandelbrot (1964) employed as an approximate model for nerve membrane potential. Roughly speaking, the following correspondences prevail between the original random-walk and its smoothed version. If excitation and inhibition arrive at the occurrence times of jumps in the two independent Poisson processes N_E and N_I, with mean rates λ_E and λ_I, and each excitatory input causes the depolarization $V(t)$ to jump up by $a_E \geq 0$ whereas each inhibitory input causes $V(t)$ to jump down by $a_I \geq 0$ units, then

$$V(t) = a_E N_E(t) - a_I N_I(t), \qquad t \geq 0. \tag{9.81}$$

To obtain an approximation to V with continuous sample paths, we use a Wiener process with drift that has the same mean and variance as V. Thus in (9.77) we put

$$\mu = a_E \lambda_E - a_I \lambda_I, \tag{9.82}$$

$$\sigma = \sqrt{a_E^2 \lambda_E + a_I^2 \lambda_I}. \tag{9.83}$$

We have, of course, left something out as we have not performed any limiting operations on the jump amplitudes or the mean rates to obtain the continuous process from the discontinuous one. We have made what has been called the usual diffusion approximation (Walsh 1981a) in the hope that the original process and the approximating smooth process go to the same places at about the same times.

9.6.1   First-passage time to threshold

We again endow our model with a threshold condition, namely, that an action potential occurs when the approximating process X reaches the level θ, assumed constant, for the first time. The firing time is then the random variable

$$T_\theta = \inf\{t \mid X(t) = \theta\}, \qquad X(0) = x_0 < \theta. \tag{9.84}$$

The density of T_θ can be found in closed form, first by the method of images when there is no drift, and second by the renewal-equation approach when drift is present.

(A) *The drift* $\mu = 0$: *method of images*
We use essentially the same argument as for the symmetric randomized random walk. The *transition probability density* function of X is defined as

$$p(x, t \mid x_0) = d/dx \, \Pr\{X(t) \le x \mid X(0) = x_0\}, \tag{9.85}$$

and for the unrestricted process this is given by (9.80). We have, roughly speaking,

$$\Pr\{X(t) \in (x, x + dx] \mid X(0) = x_0\} \simeq p(x, t \mid x_0)\, dx. \tag{9.86}$$

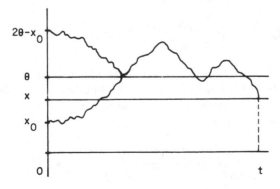

Figure 9.13. Paths of X for the method of images.

Referring to Figure 9.13, consider a path starting at x_0 and touching or rising above the level θ in $(0, t)$. For each such path there is one from the image point $2\theta - x_0$ to x. Hence

$$\Pr\{ X(t) \in (x, x + dx] \text{ and } X(t') < \theta \text{ for all } t' \in (0,t) | X(0) = x_0 \}.$$

$$= [p(x, t|x_0) - p(x, t|2\theta - x_0)] \, dx. \tag{9.87}$$

Integrating over all x less than θ and noting that if X has stayed below θ up to t, then $T_\theta > t$,

$$\Pr\{ T_\theta > t \} = \Pr\left\{ \max_{0 \le t' \le t} X(t') < \theta | X(0) = x_0 \right\}$$

$$= \int_{-\infty}^{\theta} [p(x, t|x_0) - p(x, t|2\theta - x_0)] \, dx. \tag{9.88}$$

Now (again roughly speaking) transitions from $2\theta - x_0$ to x occur with the same probabilities as transitions from x_0 to $2\theta - x$, so $p(x, t|2\theta - x_0) = p(2\theta - x, t|x_0)$. Thus, since $\Pr\{ T_\theta > t \} = 1 - \Pr\{ T_\theta \le t \}$, $t > 0$,

$$\Pr\{ T_\theta \le t \} = 1 - \int_{-\infty}^{\theta} [p(x, t|x_0) - p(2\theta - x, t|x_0)] \, dx. \tag{9.89}$$

Changing variables in the second integral yields

$$\Pr\{ T_\theta \le t \} = 2 \int_{\theta}^{\infty} p(x, t|x_0) \, dx$$

$$= \sqrt{\frac{2}{\pi\sigma^2 t}} \int_{\theta}^{\infty} \exp\left\{ -\frac{(x - x_0)^2}{2\sigma^2 t} \right\} \, dx. \tag{9.90}$$

With the additional change of variable $z = (x - x_0)/\sigma\sqrt{t}$, we get

$$\Pr\{ T_\theta \le t \} = \sqrt{\frac{2}{\pi}} \int_{(\theta - x_0)/\sigma\sqrt{t}}^{\infty} e^{-z^2/2} \, dz. \tag{9.91}$$

But the distribution function of T_θ is $F_\theta(t) = \Pr\{ T_\theta \le t \}$, and its derivative is the density f_θ of T_θ. Differentiating (9.91), we obtain the following expression for the *density of the first-passage time to level θ for the driftless Wiener Process*:

$$f_\theta(t) = \frac{\theta - x_0}{\sqrt{2\pi\sigma^2 t^3}} \exp\left\{ -\frac{(\theta - x_0)^2}{2\sigma^2 t} \right\}, \qquad t > 0, \theta > x_0. \tag{9.92}$$

Letting $t \to \infty$ in (9.91), we see immediately that passage to level θ is certain.

(B) *The general case including $\mu \neq 0$: renewal equation*
Using the same argument as for the general randomized random walk, we obtain the renewal equation

$$p(x, t|x_0) = \int_0^t f_\theta(t') p(x, t - t'|\theta) \, dt', \qquad x > \theta. \qquad (9.93)$$

Taking Laplace transforms and rearranging gives

$$f_{\theta,L}(s) = \frac{p_L(x, s|x_0)}{p_L(x, s|\theta)}. \qquad (9.94)$$

With the aid of the standard transform (Abramowitz and Stegun 1965, page 1026),

$$\mathscr{L}\left\{ \frac{1}{\sqrt{\pi t}} \exp\left(-\frac{k^2}{4t} \right) \right\} = \frac{1}{\sqrt{s}} e^{-k\sqrt{s}}, \qquad k \geq 0, \qquad (9.95)$$

we find

$$p_L(x, s|x_0) = \frac{1}{\sqrt{2\sigma^2}} \exp\left[\frac{\mu(x - x_0)}{\sigma^2} \right] \frac{\exp\left[-\frac{(x - x_0)}{\sigma} \sqrt{2(s - c)} \right]}{\sqrt{s - c}}. \qquad (9.96)$$

where $c = -\mu^2/2\sigma^2$. It is left as an exercise to show that the *Laplace transform of the first-passage time density* is

$$f_{\theta, L}(s) = \exp\left\{ \frac{(\theta - x_0)}{\sigma^2} \left(\mu - \sqrt{\mu^2 + 2\sigma^2 s} \right) \right\}. \qquad (9.97)$$

The inversion of this transform is facilitated by another standard result (Abramowitz and Stegun 1965, page 1026),

$$\mathscr{L}\left\{ \frac{k}{2\sqrt{\pi t^3}} \exp\left(-\frac{k^2}{4t} \right) \right\} = \exp(-k\sqrt{s}), \qquad k \geq 0, \qquad (9.98)$$

which yields the *inverse Gaussian* density:

$$f_\theta(t) = \frac{\theta - x_0}{\sqrt{2\pi\sigma^2 t^3}} \exp\left[-\frac{(\theta - x_0 - \mu t)^2}{2\sigma^2 t} \right], \qquad t > 0, \, \theta > x_0. \qquad (9.99)$$

Moments of the firing time

We may find the probability that X ever reaches θ by utilizing the relation

$$f_{\theta, L}(0) = \int_0^\infty f_\theta(t)\, dt = \Pr\{T_\theta < \infty\}. \tag{9.100}$$

From (9.97)

$$f_{\theta, L}(0) = \exp\left[\frac{(\theta - x_0)}{\sigma^2}(\mu - |\mu|)\right], \tag{9.101}$$

since $\sqrt{\mu^2} = |\mu|$ must be nonnegative. Since $|\mu| = \mu$ if $\mu \geq 0$ and $|\mu| = -\mu$ if $\mu < 0$, we have

$$\Pr\{T_\theta < \infty\} = \begin{cases} 1, & \mu \geq 0, \\ \exp\left[-\dfrac{2|\mu|(\theta - x_0)}{\sigma^2}\right], & \mu < 0. \end{cases} \tag{9.102}$$

Thus, if the drift is zero or toward the threshold, an action potential is generated in a finite time with probability one. On the other hand, as with the random-walk model, if the drift is away from the barrier so that $a_E\lambda_E < a_I\lambda_I$, there is probability $1 - \exp[-2|\mu|(\theta - x_0)/\sigma^2]$ that no action potential is ever generated (an eternally silent cell).

In the case $\mu \geq 0$ the mean waiting time for the occurrence of an action potential can be found from

$$E[T_\theta] = -\left.\frac{df_{\theta, L}(s)}{ds}\right|_{s=0}, \tag{9.103}$$

and the second moment can be found from

$$E[T_\theta^2] = \left.\frac{d^2 f_{\theta, L}(s)}{ds^2}\right|_{s=0}. \tag{9.104}$$

It is an exercise to show that

$$E[T_\theta] = \frac{\theta - x_0}{\mu}, \tag{9.105}$$

$$\mathrm{Var}[T_\theta] = \frac{(\theta - x_0)\sigma^2}{\mu^3}, \qquad \mu > 0,\ \theta \geq x_0. \tag{9.106}$$

When $\mu = 0$ the first- and higher-order moments of T_θ are infinite, as they must also be when $\mu < 0$.

In terms of the original physiological parameters of the model, assuming the initial value of the membrane potential is resting level, the mean ISI is

$$E[T_\theta] = \frac{\theta}{a_E \lambda_E - a_I \lambda_I}, \tag{9.107}$$

and the variance is

$$\text{Var}[T_\theta] = \frac{\theta(a_E^2 \lambda_E + a_I^2 \lambda_I)}{(a_E \lambda_E - a_I \lambda_I)^3}, \qquad a_E \lambda_E > a_I \lambda_I. \tag{9.108}$$

This gives a coefficient of variation of

$$\text{CV}[T_\theta] = \left(\frac{a_E^2 \lambda_E + a_I^2 \lambda_I}{2\theta(a_E \lambda_E - a_I \lambda_I)^3} \right)^{1/2}. \tag{9.109}$$

Again, for fixed values of the remaining parameters, the coefficient of variation of the ISI is inversely proportional to the square root of the threshold.

A numerical example
   With time in units of the membrane time constant and voltages in millivolts, we will find the mean, variance, coefficient of variation, and density of the firing time for the following parameter values: $\theta = 10\,\text{mV}$, $a_E = a_I = 1\,\text{mV}$, $\lambda_E = 2.5$, and $\lambda_I = 0.5$. Then

$$E[T_\theta] = 5,$$

$$\text{Var}[T_\theta] = \tfrac{15}{8},$$

$$\text{CV}[T_\theta] = 0.27.$$

The density of T_θ is sketched in Figure 9.14.

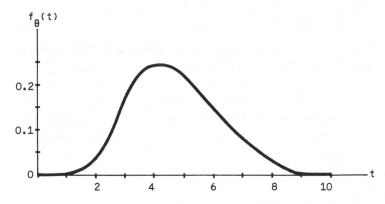

Figure 9.14. Density of the firing time for the Wiener process with drift model with parameters as given in the text.

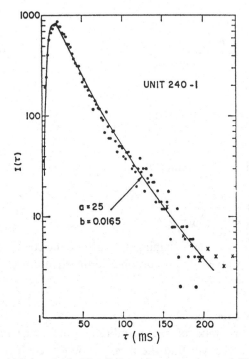

Figure 9.15. The fitting of an experimental ISI histogram to the first-passage time density of a Wiener process with drift. [From Gerstein and Mandelbrot (1964). Reproduced from *The Biophysical Journal* by copyright permission of The Biophysical Society.]

Although the Wiener process with drift has a poor physiological foundation as a model for nerve-cell membrane potential, formula (9.99) for the first-passage time density has been successfully fitted to the ISI histograms of real neurons. For example, Gerstein and Mandelbrot (1964) put $f_\theta(t)$ in the form

$$f_\theta(t) = Kt^{-3/2}\exp(-a/t - bt), \tag{9.110}$$

regarding a and b as parameters with K determined by the normalization condition $\int_0^\infty f_\theta(t)\,dt = 1$. One case of their fitting procedure is shown in Figure 9.15 for a cell in the cat cochlear nucleus. Although the agreement between experimental and theoretical firing-time distributions is excellent in this and other cases, the fitting procedure is of limited use. The model itself does not incorporate the realities of nerve-cell behavior, and the parameters are not related to the physiological variables.

9.7   Markov processes

In the following few sections we will need some basic results from the theory of *Markov processes*. This section contains a brief review of the pertinent results. Technicalities are omitted as only the applications are needed. For advanced mathematical details and foundations see such books as those of Dynkin (1965), Breiman (1968), and Gihman and Skorohod (1972, 1974, 1975, 1976).

To say that $\{X(t), t \geq 0\} \doteq X$ is a Markov process is to say that if X is known at time s, then the probabilities that X takes on its various possible values at any future time are completely determined.

Transition probability functions

Markov processes are characterized by their *transition probability functions*. Let $\{X(t), t \geq 0\}$ be a Markov process in continuous time. Suppose at time s, X is known to have the value x. We call

$$P(y, t|x, s) = \Pr\{X(t) \leq y | X(s) = x\}, \qquad s \leq t, \quad (9.111)$$

the *transition probability distribution function*.

If P is differentiable in y, its derivative is the *transition probability density function*

$$p(y, t|x, s) = \partial P/\partial y(y, t|x, s). \quad (9.112)$$

Processes for which the transition probabilities depend only on *time differences* $t - s$ are called *temporally homogeneous*. All of the random processes considered thus far in this chapter, being processes with independent increments, are Markov processes and are also temporally homogeneous. We will only be concerned in the sequel with temporally homogeneous Markov processes.

Example

The Wiener process with drift parameter μ and variance parameter σ^2 considered in the previous section has the transition probability distribution function

$$P(y, t|x, s) = \frac{1}{\sqrt{2\pi\sigma^2(t-s)}} \int_{-\infty}^{y} \exp\left[-\frac{(z - x - \mu(t-s))}{2\sigma^2(t-s)}\right] dz,$$

$$(9.113)$$

and transition probability density function

$$p(y, t|x, s) = \frac{1}{\sqrt{2\pi\sigma^2(t-s)}} \exp\left[-\frac{(y - x - \mu(t-s))^2}{2\sigma^2(t-s)}\right].$$

$$(9.114)$$

Infinitesimal generators

Markov processes admit a characterization through operators that describe the changes in the process in small time intervals. Let X be a Markov process in continuous time. Then the *infinitesimal operator*, or *infinitesimal generator*, of X is defined for suitable functions f through

$$(\mathscr{A}f)(x) = \lim_{\Delta t \downarrow 0} \frac{E[f(X(t+\Delta t)) - f(X(t))|X(t) = x]}{\Delta t}$$

(9.115)

The infinitesimal operator of a process can be calculated from a knowledge of its transition probability function. In the theory of Markov processes it is shown how the reverse step can be carried out through the Kolmogorov partial differential equations.

Examples

(i) *Poisson process.* For the Poisson process, if $X(t) = x$, then $X(t + \Delta t) = x + 1$ with probability $\lambda \Delta t + o(\Delta t)$ and $X(t + \Delta t) = x$ with probability $1 - \lambda \Delta t + o(\Delta t)$. Thus

$$E[f(X(t+\Delta t)) - f(X(t))|X(t) = x]$$
$$= \lambda \Delta t f(x+1) + (1 - \lambda \Delta t)f(x) - f(x) + o(\Delta t).$$

Putting this in (9.114) and carrying out the limiting operation gives

$$(\mathscr{A}f)(x) = \lambda[f(x+1) - f(x)].$$

(ii) *Randomized random walk.* A similar calculation shows that for the process of Equation (9.81)

$$(\mathscr{A}f)(x) = \lambda_E f(x + a_E) + \lambda_I f(x - a_I) - (\lambda_E + \lambda_I)f(x).$$

(iii) *Wiener process with drift.* For the Wiener process with drift, use of (9.114) leads to

$$(\mathscr{A}f)(x) = \frac{\sigma^2}{2}\frac{d^2 f}{dx^2} + \mu\frac{df}{dx}.$$

Diffusion processes

Roughly speaking, diffusion processes are continuous-time Markov processes whose sample paths are continuous. Such processes are characterized by their infinitesimal mean

$$\alpha(x) = \lim_{\Delta t \downarrow 0} \frac{E[X(t+\Delta t) - X(t)|X(t) = x]}{\Delta t},$$

(9.116)

and infinitesimal variance

$$\beta^2(x) = \lim_{\Delta t \downarrow 0} \frac{\mathrm{Var}[X(t+\Delta t) - X(t) | X(t) = x]}{\Delta t} \tag{9.117}$$

The infinitesimal generator of such a process is

$$(\mathscr{A}f)(x) = \alpha(x)\frac{df}{dx} + \frac{\beta^2(x)}{2}\frac{d^2f}{dx^2}. \tag{9.118}$$

Diffusion processes can be described by their *stochastic differential equations*. The process with infinitesimal generator (9.118) has the stochastic differential

$$dX = \alpha(X)\,dt + \beta(X)\,dW, \tag{9.119}$$

where W is a standard Wiener process. The differential relation is defined by its integral version

$$X(t) = X_0 + \int_0^t \alpha(X(t'))\,dt' + \int_0^t \beta(X(t'))\,dW(t'), \tag{9.120}$$

where the initial condition is $X(0) = X_0$.

The first integral in (9.120) is a Riemann integral but the second is a *stochastic integral* with respect to W (Itô 1951; Stratonovich 1966). This integral is defined and some of its properties are given, for example, in Jaswinski (1970).

Processes with jumps

Consider a Markov random process Y that has jumps of various magnitudes. Let $\nu(t, A)$ record the number of jumps of Y up to time t that have magnitudes in the set A. Suppose for fixed A, $\nu(t, A)$ is a temporally homogeneous Poisson process with mean rate $\Pi(A)$ depending on A. Then $E[\nu(t, A)] = t\Pi(A)$. Suppose further that if A_i, $i = 1, 2, \ldots, n$, are disjoint sets, then $\nu(t, A_1)$, $\nu(t, A_2), \ldots, \nu(t, A_n)$ are mutually independent.

If we integrate over all possible jump amplitudes, we recover the original process Y,

$$Y(t) = \int_{\mathbb{R}} u\nu(t, du), \tag{9.121}$$

and the total jump rate (i.e., mean rate of jumps of all magnitudes) is

$$\Lambda = \int_{\mathbb{R}} \Pi(du). \tag{9.122}$$

The process Y is a *compound Poisson process*. If the *rate measure* Π

has a density so that $\Pi(du) = \pi(u)\,du$, then

$$\Lambda = \int_{\mathbb{R}} \pi(u)\,du. \tag{9.123}$$

Example

For the randomized random walk, jumps of magnitude $+a_E$ and $-a_I$ occur with mean rates λ_E and λ_I, respectively. Then

$$\pi(u) = \lambda_E \delta(u - a_E) + \lambda_I \delta(u + a_I)$$

and the total mean rate of jumps is

$$\Lambda = \int_{\mathbb{R}} \pi(u)\,du = \lambda_E + \lambda_I.$$

A stochastic differential equation can be written down, which describes a general Markov process with diffusion and jump components

$$dX = \alpha(X)\,dt + \beta(X)\,dW + \int_{\mathbb{R}} \gamma(X, u)\nu(dt, du). \tag{9.124}$$

Again the differential is an abbreviation for the integral equation

$$X(t) = X(0) + \int_0^t \alpha(X(t'))\,dt' + \int_0^t \beta(X(t'))\,dW(t')$$

$$+ \int_0^t \int_{\mathbb{R}} \gamma(X(t'), u)\nu(dt', du), \tag{9.125}$$

where the third integral is a stochastic integral with respect to ν. It may be shown (Gihman and Skorohod 1972) that the infinitesimal generator of the process defined by (9.125) is

$$(\mathscr{A}f)(x) = \alpha(x)\frac{df}{dx} + \frac{\beta^2(x)}{2}\frac{d^2f}{dx^2} + \int_{\mathbb{R}} f(x + \gamma(x, u))\Pi(du) - \Lambda f. \tag{9.126}$$

First-exit times

The part of the theory that concerns us most is the theory of exit times as these directly relate to the random firing of neurons. Suppose $X(0) = x$ where $a < x < b$. The *first-exit time* of X from the interval (a, b) is the random variable

$$T_{ab}(x) = \inf\{t \,|\, X(t) \notin (a, b)\}, \qquad X(0) = x \in (a. b). \tag{9.127}$$

Let the distribution function of $T_{ab}(x)$ be

$$F_{ab}(x,t) = \Pr\{T_{ab}(x) \le t\}. \tag{9.128}$$

Then (Tuckwell 1976a) F_{ab} can be found as the solution of

$$\frac{\partial F_{ab}}{\partial t} = \mathscr{A}F_{ab}, \qquad x \in (a,b), \, t > 0, \tag{9.129}$$

where $\mathscr{A}$ is a partial differential–integral operator, the infinitesimal operator of x. The initial condition is

$$F_{ab}(x,0) = \begin{cases} 0, & x \in (a,b), \\ 1, & x \notin (a,b), \end{cases} \tag{9.130}$$

and with boundary conditions

$$F_{ab}(x,t) = 1, \qquad x \notin (a,b), \, t \ge 0. \tag{9.131}$$

Differentiating F_{ab} with respect to t, we get the density of the first-exit time $f_{ab}(x,t)$, which satisfies the same equation (9.129) as F_{ab} with boundary conditions

$$f_{ab}(x,t) = \delta(t), \qquad x \notin (a,b), \tag{9.132}$$

$$f_{ab}(x,0) = 0, \qquad x \in (a,b). \tag{9.133}$$

In the case of diffusion processes, where exit from (a, b) is attained by hitting either a or b, the condition $x \notin (a,b)$ can be replaced by $x = a$ or $x = b$.

The *moments* of the first-exit time are defined through

$$\mu_{n,ab}(x) = \int_0^\infty t^n f_{ab}(x,t) \, dt, \qquad n = 0,1,2,\ldots. \tag{9.134}$$

These satisfy the recursion system of equations

$$\mathscr{A}\mu_{n,ab} = -n\mu_{n-1,ab}, \qquad x \in (a,b). \tag{9.135}$$

When $n = 0$, one obtains the probability $\mu_{0,ab}$ that X ever leaves (a,b) and for this the boundary condition is

$$\mu_{0,ab}(x) = 1, \qquad x \notin (a,b). \tag{9.136}$$

For the first- and higher-order moments of $T_{ab}(x)$, the boundary conditions are

$$\mu_{n,ab}(x) = 0, \qquad x \notin (a,b), \, n = 1,2,\ldots, \tag{9.137}$$

and $\mu_{n,ab}$ is bounded on (a,b), if (a,b) is a finite interval. Exit times from intervals such as $(-\infty, b)$ are obtained by letting $a \to -\infty$ in the results for finite intervals.

9.8 Stein's model

Stein (1965) proposed a model for nerve-cell activity in the presence of random synaptic inputs. He extended the random-walk model of Section 9.5 by including the exponential decay of the membrane potential between inputs. This is a stochastic version of the Lapicque model of Chapter 3, and, in fact, the EPSPs and IPSPs are of the kind sketched in Figure 3.5. For subthreshold depolarizations $\{X(t), t \geq 0\}$, we now have the *stochastic differential* equation

$$dX = -X \, dt + \int_{\mathbb{R}} u \nu (dt, du), \qquad t > 0, \; X < \theta, \qquad (9.138)$$

where the mean-rate function is

$$\pi(u) = \lambda_E \delta(u - a_E) + \lambda_I \delta(u + a_I), \qquad (9.139)$$

with $a_E, a_I \geq 0$ and $X(0) = x < \theta$. Alternatively, we may write, somewhat heuristically,

$$\frac{dX}{dt} = -X + a_E \frac{dN_E}{dt} - a_I \frac{dN_I}{dt}, \qquad (9.140)$$

where N_E and N_I are independent simple Poisson processes with mean rates λ_E and λ_I. The trajectories of N_E and N_I have discontinuities of $+1$ each time an excitatory or inhibitory input arrives. Hence the derivatives dN_E/dt and dN_I/dt, which appear in (9.140), consist of a collection of delta functions concentrated at the random arrival times of the synaptic inputs. When an excitatory input arrives, X will jump by $+a_E$, whereas when an inhibitory input arrives X will jump by $-a_I$. Note that in (9.140) time is measured in units of the membrane time constant so that the coefficient of $-X$ is unity. Sample paths for X will appear as sketched in Figure 9.16.

The mean and variance of the depolarization in the absence of a threshold

To find the mean and variance of $X(t)$ when there is no imposed threshold condition, we set (Tuckwell 1977)

$$Y(t) = e^t X(t), \qquad (9.141)$$

in which case Y satisfies

$$\frac{dY}{dt} = e^t \left(a_E \frac{dN_E}{dt} - a_I \frac{dN_I}{dt} \right), \qquad t > 0, \qquad (9.142)$$

with $Y(0) = X(0)$. Integrating this equation gives

$$Y(t) = Y(0) + \int_0^t e^u \left(a_E \, dN_E(u) - a_I \, dN_I(u) \right). \qquad (9.143)$$

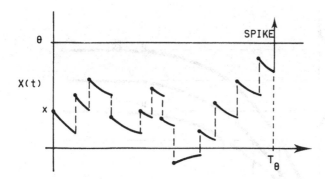

Figure 9.16. A representative sample path for the depolarization in Stein's model. An action potential is generated when X reaches or exceeds the threshold θ for the first time.

Taking expectations gives

$$E[Y(t)] = Y(0) + \int_0^t e^u (a_E \lambda_E - a_I \lambda_I)\, du$$

$$= Y(0) + (a_E \lambda_E - a_I \lambda_I)(e^t - 1). \qquad (9.144)$$

The variance is

$$\mathrm{Var}[Y(t)] = \int_0^t e^{2u}(a_E^2 \lambda_E + a_I^2 \lambda_I)\, du = \frac{(a_E^2 \lambda_E + a_I^2 \lambda_I)}{2}(e^{2t} - 1). \qquad (9.145)$$

It follows that since $E[X(t)] = e^{-t} E[Y(t)]$ and $\mathrm{Var}[X(t)] = e^{2t} \mathrm{Var}[Y(t)]$, that the *mean and variance of the depolarization in Stein's model* are

$$E[X(t)] = X(0)e^{-t} + (a_E \lambda_E - a_I \lambda_I)(1 - e^{-t}), \qquad (9.146)$$

$$\mathrm{Var}[X(t)] = \tfrac{1}{2}(a_E^2 \lambda_E + a_I^2 \lambda_I)(1 - e^{-2t}). \qquad (9.147)$$

Note that as $t \to \infty$, both the mean and the variance approach *steady-state values* in the absence of a threshold. This is in contrast with the models of the previous sections, where in the absence of a threshold the moments (usually) become unbounded as $t \to \infty$. In the case of n independent Poisson inputs with mean rates λ_i and with corresponding postsynaptic potential amplitudes a_i, the mean and variance of the depolarization are

$$E[X(t)] = X(0)e^{-t} + (1 - e^{-t}) \sum_{i=1}^n a_i \lambda_i \xrightarrow[t \to \infty]{} \sum_{i=1}^n a_i \lambda_i, \qquad (9.148)$$

$$\mathrm{Var}[X(t)] = \tfrac{1}{2}(1 - e^{-2t}) \sum_{i=1}^n a_i^2 \lambda_i \xrightarrow[t \to \infty]{} \tfrac{1}{2} \sum_{i=1}^n a_i^2 \lambda_i. \qquad (9.149)$$

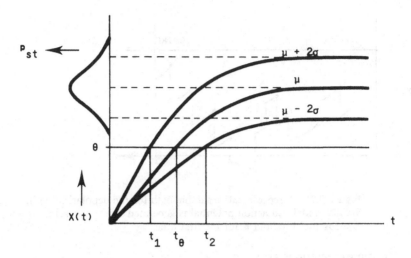

Figure 9.17. The time courses of μ_x and $\mu_x \pm 2\sigma_x$ in relation to threshold and the stationary density.

9.8.1   The interspike interval (ISI)

If there is just one excitatory input with $a_E \geq \theta$, then the occurrence of one synaptic input will cause the cell to fire. As we have seen, in this case the ISI is exponentially distributed with mean $1/\lambda_E$. This case is hereafter excluded.

First approximations

Assume a cell is initially at rest so that $X(0) = 0$. Consider Figure 9.17, where the mean depolarization $\mu_X(t)$ and the mean plus and minus 2 standard deviations $\mu_X(t) \pm 2\sigma_X(t)$, are shown plotted against time. $X(t)$ is mostly concentrated within the envelope bounded by $\mu_X(t) \pm 2\sigma_X(t)$. As $t \to \infty$ the distribution of X approaches a *stationary* (*time-invariant*) distribution. To the left of the vertical axis is sketched the stationary density p_{st}, which has mean $\sum a_i \lambda_i$ and standard deviation $(\sum a_i^2 \lambda_i / 2)^{1/2}$. We consider two cases.

(a) If $\mu_X(\infty) = \sum a_i \lambda_i > \theta$, an approximate value of the time of firing is that at which $\mu_X(t)$ reaches θ, shown as t_θ. This will be

$$t_\theta = -\ln\left[1 - \frac{\theta}{\sum a_i \lambda_i} \right], \qquad \sum a_i \lambda_i > \theta. \qquad (9.150)$$

The smaller the variance of $X(t)$, the better we expect this approximation to be, since in the limiting case of zero variability it is exact. In Figure 9.17 we have also indicated the times t_1 and t_2 at which $\mu_X(t) + 2\sigma_X(t)$ and $\mu_X(t) - 2\sigma_X(t)$ reach θ. We expect (9.150) to be a good approximation if t_1 and t_2 do not differ very much from t_θ.

That is, t_θ will be a good approximation if the following conditions are simultaneously met:

$$\sum_{i=1}^{n} a_i \lambda_i > \theta, \qquad 1 - \frac{t_1}{t_\theta} \ll 1, \qquad \frac{t_2}{t_\theta} - 1 \ll 1. \qquad (9.151)$$

The values of t_1 and t_2 can be found explicitly.

(b) If $\mu_X(\infty) = \sum a_i \lambda_i < \theta$, we cannot ever estimate $E[T_\theta]$ by the above method. However, if the variance of X is not large enough to carry X past θ in a small enough time interval, we expect the ISI to be very large. A rough estimate of when this situation arises is $\mu_X(\infty) + 2\sigma_X(\infty) < \theta$. That is, when the following conditions are satisfied, we expect the cell to be practically silent:

$$\sum_{i=1}^{n} a_i \lambda_i < \theta, \qquad \sum_{i=1}^{n} a_i \lambda_i + \left(2 \sum_{i=1}^{n} a_i^2 \lambda_i \right)^{1/2} < \theta. \qquad (9.152)$$

The moments of the firing time

From (9.126), (9.138), and (9.139) we find that the infinitesimal generator of the depolarization X in Stein's model for the case of a single excitatory and single inhibitory input is given by

$$(\mathscr{A}f)(x) = -x\frac{df}{dx} + \lambda_E f(x + a_E) + \lambda_I f(x - a_I)$$

$$- (\lambda_E + \lambda_I)f(x). \qquad (9.153)$$

Hence, from (9.135), the equations for the moments of the first-exit time of X from the interval (a, θ) can be found. We drop the a, θ subscripts and have

$$-x\frac{d\mu_n}{dx} + \lambda_E \mu_n(x + a_E) + \lambda_I \mu_n(x - a_I) - (\lambda_E + \lambda_I)\mu_n(x)$$

$$= -n\mu_{n-1}(x), \qquad x \in (a, \theta), \; n = 0, 1, 2 \ldots. \qquad (9.154)$$

This is a recursion system of differential-difference equations, which can be solved exactly in a few simple cases (Tuckwell, 1975). In general, numerical methods are required for their solution. When there is excitation only and we are only interested in threshold crossings from rest or an initially depolarized state, the equations may be solved on $[0, \theta)$, but when there is inhibition as well, we must consider (9.154) on $(-\infty, \theta)$. In all cases we have to solve the equations on their entire domain even if we are only interested in the solutions at $x = 0$.

(A) *Excitation only.*   If there is excitation only (9.154) becomes

$$-x\frac{d\mu_n}{dx} + \lambda_E[\mu_n(x + a_E) - \mu_n(x)] = -n\mu_{n-1}(x), \qquad x \in [0, \theta).$$

$$(9.155)$$

The solution of the equation with $n = 0$, with boundary data $\mu_0 = 1$ for $x \notin [0, \theta)$ is seen to be

$$\mu_0(x) = 1. \qquad (9.156)$$

That is,

$$\Pr\{T_\theta < \infty\} = 1. \qquad (9.157)$$

The equations for the first- and higher-order moments can be solved exactly in a few cases by the *method of steps*.

Examples

(i) Let $a_E = 1$, $\lambda_E = 1$, and $\theta = 2$. Consider the equation for the mean first-exit time, now denoted by $f(x)$,

$$-x\frac{df}{dx} + f(x + 1) - f(x) = -1, \qquad x \in [0, 2). \qquad (9.158)$$

We will solve this equation on $(0, 2)$ and then utilize

$$f(0) = 1 + f(1),$$

to obtain $f(0)$. The physical basis for this condition is that the mean time to the first jump is unity and this takes the process to $x = 1$ from which the mean exit time is $f(1)$.

The boundary condition is $f(x) = 0$ for $x \geq 2$ and we impose the requirements that f is continuous and bounded on $(0, 2)$. Continuity is required because the mean first-exit times from neighboring points must be close. We now employ the boundary condition to convert the differential-difference equation to two coupled differential equations.

On $[1, 2)$ let the solution be f_1 and on $(0, 1)$ let the solution be f_2. For $x \in [1, 2)$ we have $x + 1 \geq 2$ so the equation for f_1 is

$$\frac{df_1}{dx} + \frac{f_1}{x} = \frac{1}{x}. \qquad (9.159)$$

Hence (see Section 2.4)

$$f_1(x) = 1 + \frac{c_1}{x}, \qquad x \in [1, 2), \qquad (9.160)$$

where c_1 is a constant to be determined.

When $x \in (0, 1)$ we have $x + 1 \in (1, 2)$ so

$$\frac{df_2}{dx} + \frac{f_2}{x} = \frac{2}{x} + \frac{c_1}{x(x + 1)}, \qquad x \in (0, 1). \qquad (9.161)$$

The general solution of this equation is

$$f_2(x) = 2 + \frac{c_1 \ln(x+1)}{x} + \frac{c_2}{x}. \tag{9.162}$$

At $x = 1$ continuity of f demands that $f_2(1^-) = f_1(1)$, which yields

$$1 + c_1 = 2 + c_1 \ln 2 + c_2. \tag{9.163}$$

Consider now the behavior of $f_2(x)$ as $x \to 0$. Since

$$\ln(x+1) = x - x^2/2 + \cdots,$$

the second term is bounded as $x \to 0$. However, $c_2/x \to \infty$ as $x \to 0$ for any nonzero c_2. Boundedness therefore requires $c_2 = 0$. We then get

$$c_1 = \frac{1}{1 - \ln 2}. \tag{9.164}$$

The solution of (9.158) is, in the original notation,

$$\mu_1(x) = \begin{cases} 2 + (1 - \ln 2)^{-1} \dfrac{\ln(x+1)}{x}, & 0 < x < 1, \\[2ex] 1 + \dfrac{(1 - \ln 2)^{-1}}{x}, & 1 \le x < 2. \end{cases} \tag{9.165}$$

In addition, using $f(0) = 1 + f(1)$, we obtain for $\mu_1(0)$

$$E[T_\theta] = 2 + \frac{1}{1 - \ln 2}. \tag{9.166}$$

The solution is sketched in Figure 9.18.

If we use a time constant appropriate for cat spinal motoneurons, this corresponds roughly to an output frequency of 32 s^{-1}. It is important to note, however, that use of the reciprocal of $E[T_\theta]$ to compute a mean firing rate is quite crude, since $1/E[T_\theta] \neq E[1/T_\theta]$.

There are two points to make.

1. If there were no exponential decay, we would have $E[T_\theta] = 2$, so the extra term $(1 - \ln 2)^{-1} \approx 3.26$ reflects the contributions from the decay between EPSPs.

2. If the input were constant with the same mean value so that $dX/dt = -X + a_E \lambda_E$, the cell would never fire since the asymptotic $(t \to \infty)$ depolarization would be less than θ. Thus the effect of variability of the input is to decrease the mean ISI.

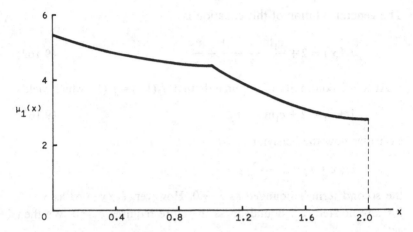

Figure 9.18. The expected time $\mu_1(x)$ at which the depolarization $X(t)$, with $X(0) = x$, first exceeds threshold for Stein's model neuron with excitation only. Parameter values $\theta = 2$, $\lambda_E = 1$, and $a_E = 1$.

(ii) Let $1 < \theta \leq 2$. Exact solutions for the first and second moment of the ISI can be found for excitation only when $a_E = 1$, and $\lambda_E = n$ or $1/n$, $n = 1, 2, \ldots$. Note that setting $a_E = 1$ means a reinterpretation of θ as the ratio of the threshold to the EPSP amplitude. In general, we have the following equation for the mean exit time:

$$-x\frac{df}{dx} + \lambda_E[f(x+1) - f(x)] = -1, \qquad x \in [0, \theta).$$

$$(9.167)$$

Denoting the second moment by $g(x)$, we have

$$-x\frac{dg}{dx} + \lambda_E[g(x+1) - g(x)] = -2f(x), \qquad x \in [0, \theta).$$

$$(9.168)$$

The following expressions are obtained for the first and second moments of the time to reach threshold from resting level:

$$E[T_\theta] = \frac{2}{\lambda_E} + \alpha, \qquad (9.169)$$

$$E[T_\theta^2] = \frac{2}{\lambda_E^2} + \frac{2E[T_\theta]}{\lambda_E} + \frac{4(\theta-1)^{\lambda_E} + 2\alpha[\lambda_E I_2 - \ln(\theta-1)]}{1 - \lambda_E I_1},$$

$$(9.170)$$

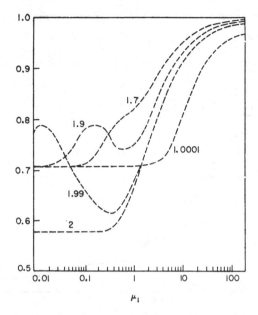

μ_1

Figure 9.19. Coefficient of variation versus mean interval for Stein's model with excitation only for selected values of the threshold θ, which are indicated on each curve. [From Tuckwell and Richter (1978). Reproduced with the permission of Academic and the authors.]

where

$$\alpha = \frac{(\theta - 1)^{\lambda_E}}{\lambda_E (1 - \lambda_E I_1)}, \tag{9.171}$$

$$I_1 = \left(1 - \frac{1}{\theta}\right)^{\lambda_E} \sum_{j=0}^{\infty} \frac{(1 - 1/\theta)^j}{j + \lambda_E}, \tag{9.172}$$

$$I_2 = I_1 \ln \theta + \left(1 - \frac{1}{\theta}\right)^{\lambda_E} \sum_{j=0}^{\infty} \frac{(1 - 1/\theta)^j}{(j + \lambda_E)^2}. \tag{9.173}$$

From these formulas the coefficient of variation of the ISI may be found for various input frequencies. Figure 9.19 shows a semilogarithmic plot of $\mathrm{CV}[T_\theta]$ versus $E[T_\theta]$. It can be seen that the coefficient of variation, which is used to quantify the regularity of a spike train, is not always a monotonically increasing function of the mean interval and that sometimes $\theta_1 > \theta_2$ does not imply $\mathrm{CV}[T_{\theta_1}] < \mathrm{CV}[T_{\theta_2}]$ for all values of the mean interval. These facts were not discernible from the computer-simulation studies of Stein (1967a). See Enright (1967) and Tuckwell and Richter (1978) for further discussion.

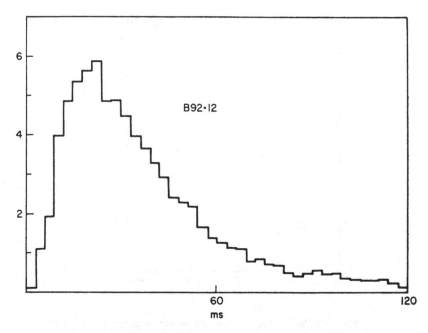

Figure 9.20. The ISI histogram of a cell in the cat cochlear nucleus. [From Tuckwell and Richter (1978); data supplied by Terry Bourk. Reproduced with the permission of Academic and the authors.]

For $\theta \le 5$ exact or extrapolated mean ISIs have been calculated for Stein's model for $\lambda_E = n$ and $\lambda_E = 1/n$, $n = 1, 2, 3$. These have been compared with the experimental results of Redman and Lampard (1968) on Poisson stimulation of cat spinal motoneurons. Given the uncertainties in the parameters and the actual input to the motoneurons, reasonable agreement between theory and experiment was obtained (Tuckwell 1976b). In a similar study, the output-versus-input-frequency relations were obtained for dorsalspinocerebellar tract cells (Tuckwell 1976c); see also Walloe, Jansen, and Nygaard (1969).

For larger values of θ numerical methods of solution have been employed to determine the first three moments of the ISI (Tuckwell and Richter 1978). Such results were used to estimate the time constant, input rate of excitation, and threshold for cells of the cat cochlear nucleus by the method of moments, complementing an earlier study by Molnar and Pfeiffer (1968). The experimental histogram of ISIs for one of these cells in its spontaneously active state is shown in Figure 9.20 The estimated parameters for this cell were: time constant, 9.1 ms; input rate of excitation, 352 s^{-1}; and threshold to EPSP amplitude, 4.9. Despite the uncertainties and limitations of the model, these estimated values are all physiologically reasonable.

Furthermore, on the basis of the known firing rates of the auditory nerve fibers that synapse with these cells, it was deduced that the number of such fibers connecting with the cell under consideration was 9. Thus the possible usefulness of the ISI histogram in revealing neurophysiological and neuroanatomical information not otherwise accessible to the experimenter was in part indicated.

(B) *Excitation with inhibition.* When there is inhibition as well as excitation, the moment equations are given by (9.154). These are differential-difference equations with both forward and backward differences and the method of steps fails. There are no known standard methods of solution for these equations.

Bearing in mind the results for the randomized random walk, our first concern is whether the depolarization will reach θ in a finite time with probability one, and if this depends on the strength of the inhibition relative to the excitation. We find that as long as $\lambda_E > 0$ and $a_E > 0$, the solution of (9.154) with $n = 0$, which satisfies the boundary condition $\mu_0(x) = 1$ for $x \notin (a, \theta)$, is, in fact, $\mu_0(x) = 1$. Hence $\Pr\{T_\theta < \infty\} = 1$ and $X(t)$ is guaranteed to reach θ in a finite time.

A method has been devised (Cope and Tuckwell 1979) for solving the equation for the mean first-passage time to θ. When, for convenience, $a_E = a_I = 1$, we have

$$-x\,df/dx + \lambda_E f(x+1) + \lambda_I f(x-1) - (\lambda_E + \lambda_I)f(x) = -1,$$
$$x < \theta. \quad (9.174)$$

It was argued that at large negative x, $f(x)$, $f(x+1)$, and $f(x-1)$ should be close so that, asymptotically,

$$-x\,df/dx \simeq -1, \qquad x \to -\infty. \quad (9.175)$$

Thus for large negative x the mean first-exit time should be of the form

$$f(x) \underset{x \to -\infty}{\sim} \ln|x| + C, \quad (9.176)$$

where C is a constant to be determined. One is led, therefore, to try a solution of the form

$$f(x) \underset{x \to -\infty}{\sim} \ln|x| + C + \sum_{n=1}^{\infty} A_n x^{-n}. \quad (9.177)$$

The coefficients A_n are found by substituting (9.177) in (9.174). The asymptotic solution can be continued toward θ, since, if f is known on two neighboring unit intervals (9.174) becomes an algebraic equation. Finally, the value of C is found from the boundary condition at $x \geq \theta$. Full details are in Cope and Tuckwell (1979).

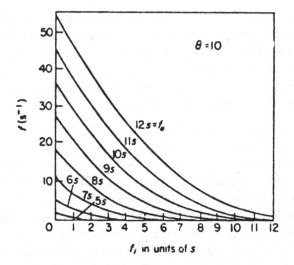

Figure 9.21. Output frequency f for Stein's model neuron as a function of frequency of inhibition for various frequencies of excitation. $s = 1/$ time constant and time constant $= 12$ ms. The threshold is 10 and an absolute refractory period of 1 ms is assumed. [From Cope and Tuckwell (1979). Reproduced with the permission of Academic and the authors.]

One set of results obtained by the above method are shown in Figure 9.21. The reciprocal of the mean interspike interval is shown as a function of mean input frequency of inhibition for $\theta = 10$ and various frequencies of excitation. A time constant of 12 ms was assumed, which is an appropriate value for fast pyramidal tract cells of the cat (Calvin and Sypert 1976). The output frequencies for zero inhibitory input frequencies agreed well with those obtained by other methods. It can be seen that the effects of even small amounts of inhibition on firing rate can be severe.

ISI densities

One of the reasons for studying neuronal variability is that a cogent theory with corresponding results might reveal facts about nerve cells, which might otherwise be difficult, or perhaps impossible, to glean. This led Braitenberg (1965), for example, to inquire, "What can be learned from spike interval histograms about synaptic mechanisms?" He analyzed the ISI histograms of frog cerebellar Purkinje cells and presented a theoretical argument that led to a suspected threshold time course after a spike.

To provide one illustrative example, we may examine the different shapes of the ISI density for the albeit unrealistic Poisson process model. If the threshold is less than or equal to one EPSP amplitude from rest, the ISI density is exponential. If θ/a_E is between 1 and 5 or so, the density is a skewed gamma with a mode less than the mean, whereas for larger θ/a_E the density assumes a Gaussian appearance. These trends are preserved in much more realistic models as other parameters change (see Section 9.10).

Progress in obtaining ISI densities for Stein's model and models of a related nature has been slow due to the difficult mathematical problems they generate. It has been very useful to resort to computer simulation. This was first done by Stein (1965, 1967a) who found that many ISI histograms so obtained could be well approximated by gamma densities. it was also suggested that the CV-versus-mean-ISI relations could give information about threshold and time constant. Molnar and Pfeiffer (1968) obtained ISI densities for Stein's model in the case of excitation only by numerical methods and found good agreement with experimental ISI histograms of cells of the postero-ventral cochlear nucleus of the cat. Morjanoff (1971) simulated the experiments of Redman and Lampard (1968) on Poisson stimulation of cat spinal motoneurons and found that the most likely cause of frequency limiting in such cells was a decline in EPSP amplitude with increasing input frequency – a postulate supported by the exact calculations of the mean ISI (Tuckwell 1976b). In subsequent simulations it was found likely that the division of cells in the cat cerebral cortex into those with lognormal ISIs and those with non-lognormal ISIs was due to the different amounts of inhibition received by cells in each category (Tuckwell 1978a; Burns and Webb 1976). The quantitatively different effects of afterhyperpolarization and inhibition on the ISI histogram were also reported (Tuckwell 1978b). More recently, Wilbur and Rinzel (1982) have calculated accurate solutions of the equations for the ISI density by numerically inverting its Laplace transform.

9.8.2 Modifications and generalizations of Stein's model

Several modifications and generalizations of the model equation (9.140) have been proposed in a quest for more physiological realism.

(i) Random PSP amplitudes

Suppose instead of always having the same effect on the membrane potential, the EPSPs and IPSPs have random amplitudes with probability densities $\phi_E(u)$ and $\phi_I(u)$, respectively, where these densities are normalized to

$$\int_{-\infty}^{\infty} \phi_{E,I}(u)\,du = 1. \tag{9.178}$$

By definition ϕ_E must be zero for $u < 0$, and ϕ_I must be zero for $u > 0$. In Stein's original model $\phi_E = \delta(u - a_E)$ and $\phi_I = \delta(u + a_I)$, $a_E, a_I > 0$. In the general case,

$$dX = -X \, dt + \int_0^\infty u\nu(dt, du) + \int_{-\infty}^0 u\nu(dt, du). \qquad (9.179)$$

If the overall mean rates of excitation and inhibition are λ_E and λ_I, then the rate measure associated with ν has the density

$$\pi(u) = \lambda_E \phi_E(u) + \lambda_I \phi_I(u), \qquad (9.180)$$

so that still the total mean jump rate is $\Lambda = \lambda_E + \lambda_I$.

From (9.126), the infinitesimal generator of X is now given by

$$(\mathscr{A}f)(x) = -x \frac{df}{dx} + \lambda_E \int_0^\infty f(x + u)\phi_E(u) \, du$$
$$+ \lambda_I \int_{-\infty}^0 f(x + u)\phi_I(u) \, du - \Lambda f(x). \qquad (9.181)$$

The mean and variance of $X(t)$ can be found by the same method as before. From the integral representation

$$X(t) = X(0)e^{-t} + \int_0^t e^{-(t-t')}\left\{ \int_0^\infty u\nu(dt', du) + \int_{-\infty}^0 u\nu(dt', du) \right\}, \qquad (9.182)$$

we have

$$E[X(t)] = X(0)e^{-t} + e^{-t}\int_0^t e^{-t'} \, dt'\left[\lambda_E \int_0^\infty u\phi_E(u) \, du \right.$$
$$\left. + \lambda_I \int_{-\infty}^0 u\phi_I(u) \, du \right], \qquad (9.183)$$

so that the *mean depolarization* is

$$E[X(t)] = X(0)e^{-t} + (\lambda_E \mu_E + \lambda_I \mu_I)(1 - e^{-t}), \qquad (9.184)$$

where μ_E and μ_I are the mean EPSP and mean IPSP amplitudes

$$\mu_E = \int_0^\infty u\phi_E(u) \, du, \qquad \mu_I = \int_{-\infty}^0 u\phi_I(u) \, du. \qquad (9.185)$$

We also have

$$\text{Var}[X(t)] = e^{-2t}\int_0^t e^{2t'} \, dt'\left[\lambda_E \int_0^\infty u^2\phi_E(u) \, du \right.$$
$$\left. + \lambda_I \int_{-\infty}^0 u^2\phi_I(u) \, du \right], \qquad (9.186)$$

which gives the *variance of the depolarization*

$$\text{Var}[X(t)] = [\lambda_E(\mu_E^2 + \sigma_E^2) + \lambda_I(\mu_I^2 + \sigma_I^2)][1 - e^{-2t}], \qquad (9.187)$$

where σ_E^2 and σ_I^2 are the variances of the EPSP and IPSP amplitudes, respectively.

A frequently used choice of PSP amplitude distribution is zero inhibition with an EPSP that is exponentially distributed

$$\lambda_I = 0, \qquad \phi_E(u) = \frac{1}{\mu_E} e^{-u/\mu_E}, \qquad u > 0. \tag{9.188}$$

In this case the mean first-passage time to θ satisfies

$$-x\frac{df}{dx} - \lambda_E f + \frac{\lambda_E}{\mu_E}\int_0^\infty f(x+u)e^{-u/\mu_E}\,du = -1, \qquad x \in (0, \theta). \tag{9.189}$$

Losev (1975) and Tsurui and Osaki (1976) have found expressions for the mean firing time for such a PSP amplitude distribution. Losev, Shik, and Yagodnitsyn (1975) have used their results to estimate input frequency, threshold, and time constant for cells in the midbrain of the cat. Vasudevan and Vittal (1982) have found Laplace transforms of the first-passage time density when the PSPs have a Pareto distribution.

(ii) *Inclusion of reversal potentials*

As seen in Chapter 7, the EPSP and IPSP amplitudes depend on potential, both being diminished as the respective reversal potentials are approached. If the reversal potentials are included, the stochastic equation for the depolarization becomes (Tuckwell 1979a)

$$dX = -X\,dt + (V_E - X)a_E\,dN_E + (V_I - X)a_I\,dN_I, \tag{9.190}$$

where $V_I \le 0 < V_E$ are the inhibitory and excitatory reversal potentials and where now a_E and a_I are both nonnegative.

The mean depolarization in the absence of a threshold is (Tuckwell 1981b)

$$E[X(t)] = \frac{k_2}{k_1} + \left[X(0) - \frac{k_2}{k_1}\right]\exp[-k_1 t], \tag{9.190A}$$

where $k_1 = 1 + a_E\lambda_E + a_I\lambda_I$ and $k_2 = a_E\lambda_E V_E + a_I\lambda_I V_I$. The process X differs from those previously encountered because even in the absence of absorbing barriers, if $X(0) \in (V_I, V_E)$, then $X(t) \in (V_I, V_E)$ for all $t > 0$. In the notation of (9.124) we have

$$dX = -X\,dt + \int_{\mathbb{R}} \gamma(X, u)\nu(dt, du). \tag{9.191}$$

To specify γ and ν we let A_E and A_I be two disjoint intervals in $\mathbb{R}$

and let

$$\Pi(A_E) = \lambda_E, \qquad \Pi(A_I) = \lambda_I, \qquad \Lambda = \lambda_E + \lambda_I. \qquad (9.192)$$

Now, put

$$\gamma(X, u) = \begin{cases} (V_E - X)a_E, & u \in A_E, \\ (V_I - X)a_I, & u \in A_I, \\ 0, & \text{otherwise.} \end{cases} \qquad (9.193)$$

Then, from (9.126), the infinitesimal generator of X is given by

$$(\mathscr{A}f)(x) = -x\frac{df}{dx} + \int_{A_E} f(x + \gamma(x, u))\Pi(du)$$

$$+ \int_{A_I} f(x + \gamma(x, u))\Pi(du) - \Lambda f$$

$$= -x\frac{df}{dx} + \lambda_E f(x + (V_E - x)a_E)$$

$$+ \lambda_I f(x + (V_I - x)a_I) - \Lambda f. \qquad (9.194)$$

Thus the equations for the moments of the firing time can be written down using (9.135).

When there is excitation only and $\lambda_E = 1$, the mean firing time is obtained from

$$-x\frac{df}{dx} + f(x + (V_E - x)a_E) - f(x) = -1, \qquad x \in (0, \theta), \qquad (9.195)$$

with $f = 0$ for $x \notin (0, \theta)$. The method of steps can be employed to find exact solutions of this equation for small enough θ and special parameter values. If we choose $\theta = a_E V_E(2 - a_E)$, we obtain, for $f(0)$,

$$E[T_\theta] = 2 + \frac{1 - \theta/a_E V_E}{a_E - 1 + \ln(\theta/a_E V_E)}. \qquad (9.196)$$

With $V_E = 50$ and $a_E = 0.02$, θ is 1.98 and the expected interspike interval is 5.3007 as opposed to 5.0924 when the reversal potential is neglected and the EPSP amplitude is taken as that elicited at rest. When $V_E = 5$ and $a_E = 0.2$, θ is 1.9 and (9.196) gives $E[T_\theta] = 5.7698$, whereas Stein's model gives 3.9407, a difference of 31.7%.

A diffusion approximation to (9.190) has recently been analyzed (Hanson and Tuckwell 1983); see also Johannesma (1968).

Computer simulations have been carried out for (9.190) both for excitation and for excitation with inhibition (see Figure 9.22). It was

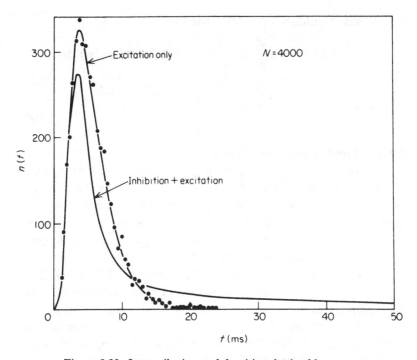

Figure 9.22. Interspike-interval densities obtained by computer simulation of solutions of (9.190). The inhibition is at one-half the intensity of the excitation. [From Tuckwell (1979a).]

found that when inhibition was excessive a coefficient of variation of T_θ greater than unity and a very long tail on the ISI density were sometimes obtained (Tuckwell 1979a). However, a comprehensive investigation by Wilbur and Rinzel (1983) using both numerical and Monte Carlo methods has cast doubt on this claim. Smith and Smith (1984) have analyzed the case of small output variability and allowed for additive white noise in (9.190).

The model including reversal potentials can be generalized to include random PSP amplitudes at fixed values of the depolarization. Then

$$\pi(u) = \begin{cases} \phi_E(u), & u > 0, \\ \phi_I(u), & u < 0, \end{cases} \tag{9.197}$$

$$\gamma(X, u) = \begin{cases} (V_E - X)u, & u > 0, \\ (V_I - X)|u|, & u < 0, \end{cases} \tag{9.198}$$

$$\Lambda = \int_0^\infty \phi_E(u)\, du + \int_{-\infty}^0 \phi_I(u)\, du, \tag{9.199}$$

and X has infinitesimal generator given by

$$(\mathscr{A}f)(x) = -x\frac{df}{dx} + \int_0^\infty f(x + (V_E - x)u)\phi_E(u)\,du$$
$$+ \int_{-\infty}^0 f(x + (V_I - x)|u|)\phi_I(u)\,du - \Lambda f.$$

(9.200)

9.9   The Ornstein–Uhlenbeck process

In the models of the previous section the membrane potential is a discontinuous function of time. The equations for the density and moments of the firing time were differential-difference or other difficult functional differential equations. Just as the Wiener process with drift can be viewed as a smoothed version of the randomized random walk, we can obtain smoothed versions of Stein's model. This leads to the *Ornstein–Uhlenbeck (1930) process* (OUP), which is discussed in many texts on probability and stochastic processes [see, for example, Cox and Miller (1965) and Breiman (1968)]. This process has arisen in several fields. In fact, in such areas as astrophysics (Chandrasekhar 1943) and electrical engineering (Stumpers 1950), the same kinds of problems have arisen as confront us in the determination of the firing times of neurons. The OUP seems to have been first mentioned in the present context by Calvin and Stevens (1965).

Our basic approach is to construct a diffusion process with the same first and second infinitesimal moments [see (9.116) and (9.117)] as the potential in Stein's model. This is called the *usual* diffusion approximation by Walsh (1981a) and it is expected to perform best when the EPSP and IPSP amplitudes are small and the input frequencies are large. The diffusion approximation therefore has infinitesimal mean

$$a_E\lambda_E - a_I\lambda_I - x \doteq \mu - x, \qquad a_E, a_I > 0, \tag{9.201}$$

and infinitesimal variance

$$a_E^2\lambda_E + a_I^2\lambda_I = \sigma^2. \tag{9.202}$$

The stochastic differential equation is thus

$$dX = (-X + \mu)\,dt + \sigma\,dW, \tag{9.203}$$

which is that of an Ornstein–Uhlenbeck process.

Mean, variance, and transition density
If we put $Y(t) = e^t X(t)$, the following equation is obtained:

$$dY = e^t(\mu\,dt + \sigma\,dW), \tag{9.204}$$

with $Y(0) = X(0)$. On integrating,

$$Y(t) = Y(0) + \mu \int_0^t e^{t'} \, dt' + \sigma \int_0^t e^{t'} \, dW(t'). \tag{9.205}$$

The mean and variance can be obtained as in the previous section:

$$E[X(t)] = X(0)e^{-t} + \mu(1 - e^{-t}) \doteq X(0)e^{-t} + m(t), \tag{9.206}$$

$$\text{Var}[X(t)] = \sigma^2(1 - e^{-2t})/2 \doteq s^2(t). \tag{9.207}$$

We observe that $X(t)$ is obtained from W by linear operations

$$X(t) = X(0)e^{-t} + \mu e^{-t} \int_0^t e^{t'} \, dt' + \sigma e^{-t} \int_0^t e^{t'} \, dW(t'). \tag{9.208}$$

Hence $X(t)$ is Gaussian and the transition probability density must be

$$p(y, t|x) = \frac{1}{\sqrt{2\pi s^2(t)}} \exp\left[\frac{-(y - xe^{-t} - m(t))^2}{2s^2(t)} \right],$$

$$y \in (-\infty, \infty). \tag{9.209}$$

Alternatively, $p(y, t|x)$ can be obtained by solving the forward Kolmogorov equation [see, for example, Gluss (1967)]. As $t \to \infty$, the distribution of $X(t)$ becomes a time-invariant Gaussian distribution with mean μ and variance $\sigma^2/2$.

First-passage time density and its Laplace transform
From (9.118) the infinitesimal generator of X is given by

$$(\mathscr{A}f)(x) = (\mu - x)\frac{df}{dx} + \frac{\sigma^2}{2}\frac{d^2f}{dx^2}. \tag{9.210}$$

Suppose now $X(0) = x \in (a, b)$. From Section 9.7 the density of the first-exit time from (a, b), denoted by $f_{ab}(x, t)$, is the solution of

$$\frac{\partial f_{ab}}{\partial t} = (\mu - x)\frac{\partial f_{ab}}{\partial x} + \frac{\sigma^2}{2}\frac{\partial^2 f_{ab}}{\partial x^2}, \qquad t > 0, \ x \in (a, b), \tag{9.211}$$

with boundary data

$$f(a, t) = f(b, t) = \delta(t), \tag{9.212}$$

and initial data

$$f(x, 0) = 0. \tag{9.213}$$

Equation (9.211) has proven difficult to solve. However, much information can be obtained from the Laplace transform

$$f_{ab,L}(x,s) = \int_0^\infty e^{-st} f_{ab}(x,t)\, dt, \qquad (9.214)$$

of f_{ab}. If we Laplace transform (9.211) and its boundary conditions, we find

$$\frac{\sigma^2}{2} \frac{d^2 f_{ab,L}}{dx^2} + (\mu - x) \frac{df_{ab,L}}{dx} - s f_{ab,L} = 0, \qquad x \in (a,b),$$

$$(9.215)$$

with boundary conditions

$$f_{ab,L}(a,s) = f_{ab,L}(b,s) = 1. \qquad (9.216)$$

The *unrestricted* (free-motion) OUP takes values in $(-\infty, \infty)$. We are interested in the time $T_\theta(x)$ taken to first reach $\theta > 0$ for an initial value $x < \theta$. The corresponding first-passage time density $f_\theta(x,t)$ is obtained from

$$f_\theta(x,t) = \lim_{a \to \infty} f_{-a,\theta}(x,t), \qquad a > 0, \qquad (9.217)$$

a similar equation applying for the Laplace transform.

The Laplace transform $f_{\theta,L}$ of f_θ has been obtained by solving (9.215) and using (9.217) for various parameter values (Siegert 1951; Roy and Smith 1969; Capocelli and Ricciardi 1971; Sugiyama, Moore, and Perkel 1970; Clay and Goel 1973; Kryukov 1976). Roy and Smith obtained $f_{\theta,L}(x,s)$ in the form

$$f_{\theta,L}(x,s) = \frac{\Psi\left(\dfrac{s}{2}, \dfrac{1}{2}; \left(\dfrac{\mu - x}{\sigma}\right)^2\right)}{\Psi\left(\dfrac{s}{2}, \dfrac{1}{2}; \left(\dfrac{\mu - \theta}{\sigma}\right)^2\right)}, \qquad (9.218)$$

where $\Psi(a,b;z)$ is the confluent hypergeometric function of the second kind (Abramowitz and Stegun 1965). This is a solution of

$$z \frac{d^2 w}{dz^2} + (b - z) \frac{dw}{dz} - aw = 0. \qquad (9.219)$$

In terms of the confluent hypergeometric function of the first kind,

$$\Phi(a,b;z) = 1 + \frac{az}{b} + \frac{a(a+1)}{b(b+1)} \frac{z^2}{2!} + \cdots$$

$$+ \frac{\Gamma(a+n)}{\Gamma(b+n)} \frac{\Gamma(b)}{\Gamma(a)} \frac{z^n}{n!} + \cdots, \qquad (9.220)$$

we have

$$\Psi(a, b; z) = \frac{\pi}{\sin \pi b} \left\{ \frac{\Phi(a, b; z)}{\Gamma(1 + a - b)\Gamma(b)} \right.$$
$$\left. - \frac{z^{1-b}\Phi(1 + a - b, 2 - b; z)}{\Gamma(a)\Gamma(2 - b)} \right\}.$$

(9.221)

Alternatively, $f_{\theta, L}(x, s)$ can be expressed in terms of parabolic cylinder functions (Abramowitz and Stegun 1965),

$$f_{\theta, L}(x, s) = \frac{\exp\left\{\left(\frac{\mu - x}{\sigma\sqrt{2}}\right)^2\right\} D_{-s}\left(\frac{\mu - x}{\sigma/\sqrt{2}}\right)}{\exp\left\{\left(\frac{\mu - \theta}{\sigma\sqrt{2}}\right)^2\right\} D_{-s}\left(\frac{\mu - \theta}{\sigma/\sqrt{2}}\right)},$$

(9.222)

which is obtained from the expression given by Sugiyama et al. (1970). The parabolic cylinder function in (9.222) is related to the confluent hypergeometric function of the first kind by

$$D_{-s}(x) = \frac{e^{-x^2/4}}{2^{s/2}\sqrt{\pi}} \left[\cos\left(\frac{\pi s}{2}\right)\Gamma\left(\frac{1}{2} - \frac{s}{2}\right)\Phi\left(\frac{s}{2}, \frac{1}{2}; \frac{x^2}{2}\right) \right.$$
$$\left. - 2^{1/2}\sin\left(\frac{\pi s}{2}\right)\Gamma\left(1 - \frac{s}{2}\right)x\Phi\left(\frac{s}{2} + \frac{1}{2}, \frac{3}{2}; \frac{x^2}{2}\right) \right].$$

(9.223)

Derivation of (9.218) and (9.222) is left as an exercise.

The inversion of $f_{\theta, L}(x, s)$ has not been performed though it is possible that numerical inversion may be useful. An alternative approach to the Laplace transform method is to utilize the absorbing barrier solution of the *forward Kolmogorov equation* satisfied by the transition probability density function $p(y, t|x)$,

$$\frac{\partial p}{\partial t} = -\frac{\partial}{\partial y}((\mu - y)p) + \frac{\sigma^2}{2}\frac{\partial^2 p}{\partial y^2}.$$

(9.224)

If there are *absorbing barriers* at a and b and $X(0) = x \in (a, b)$, then, denoting the solution of (9.224) by p_{ab}, we have (Cox and Miller 1965)

$$p_{ab}(a, t|x) = p_{ab}(b, t|x) = 0,$$

(9.225)

$$p_{ab}(y, 0|x) = \delta(x - y).$$

(9.226)

Setting $b = \theta > 0$ and letting $a \to -\infty$, we obtain the transition probability density function of the OUP in the presence of an absorbing barrier at θ, which we denote by p_θ. In terms of this absorbing barrier solution, the first-passage time density is seen to be given by

$$f_\theta(x, t) = -\frac{\partial}{\partial t} \int_{-\infty}^{\theta} p_\theta(y, t|x) \, dy. \tag{9.227}$$

Furthermore, if a *reflecting barrier* condition is imposed at a (Cox and Miller 1965)

$$\left\{ (\mu - y) p - \frac{\sigma^2}{2} \frac{\partial p}{\partial y} \right\}\Bigg|_{y=a} = 0, \tag{9.228}$$

then escape at a is impossible. Denoting the solution of (9.224) with an absorbing barrier at θ and a reflecting barrier at a by $p^*_{a\theta}$, the density of the first-passage time to θ is

$$f^*_{a\theta}(x, t) = -\frac{\partial}{\partial t} \int_{a}^{\theta} p^*_{a\theta}(y, t|x) \, dy. \tag{9.229}$$

Furthermore, from (9.224) we find, on performing the integration,

$$f^*_{a\theta}(x, t) = -\frac{\sigma^2}{2} \frac{\partial p^*_{a\theta}}{\partial y}\Bigg|_{\theta}. \tag{9.230}$$

If we let $a \to -\infty$, we recover the density of the time of first passage to θ,

$$f_\theta(x, t) = \lim_{a \to -\infty} -\frac{\sigma^2}{2} \frac{\partial p^*_{a\theta}}{\partial y}\Bigg|_{\theta}. \tag{9.231}$$

This approach does not seem to have been utilized but was mentioned by Gluss (1967) and Johannesma (1968).

The first-passage time density has been obtained by direct solution of its partial differential equation (Matsuyama, Shirai, and Akizuki 1974). The tables of Keilson and Ross (1975) may be used to find the density of T_θ for particular parameter values. Sato (1978) has given an asymptotic analysis of f_θ as $t \to \infty$.

The moments of the firing time

The moments $\mu_n(x)$, $n = 0, 1, 2, \ldots$, of the first-passage (exit) time when $X(0) = x \in (a, \theta)$ may be found from the Laplace transform of the density by means of

$$\mu_n(x)(-1)^n \frac{d^n f_L(x, s)}{ds^n}\Bigg|_{s=0}. \tag{9.232}$$

Alternatively, one may solve the recursion system obtained from (9.135) [see also Darling and Siegert (1953)]

$$\frac{\sigma^2}{2}\frac{d^2\mu_n}{dx^2} + (\mu - x)\frac{d\mu_n}{dx} = -n\mu_{n-1}, \qquad x \in (a, \theta). \quad (9.233)$$

Using the equation with $n = 0$, we may show that if $T_\theta(x)$ is time of first passage to θ from $X(0) = x < \theta$, then

$$\Pr\{T_\theta(x) < \infty\} = 1. \quad (9.234)$$

To do this we solve

$$\frac{\sigma^2}{2}\frac{d^2\mu_0}{dx^2} + (\mu - x)\frac{d\mu_0}{dx} = 0, \quad (9.235)$$

with the condition that exit occurs from (a, θ) at θ before a. That is,

$$\mu_0(a) = 0, \qquad \mu_0(\theta) = 1. \quad (9.236)$$

It is left as an exercise to show that the solution is

$$\mu_0(x) = \frac{\int_a^x \exp\left[2(y^2/2 - \mu y)/\sigma^2\right] dy}{\int_a^\theta \exp\left[2(y^2/2 - \mu y)/\sigma^2\right] dy}. \quad (9.237)$$

As $a \to -\infty$, $\mu_0(x) \to 1$. Hence (9.234) follows. Thus in the OUP model the nerve cell fires in a finite time with probability 1.

The mean time to firing for an initially resting cell was obtained from (9.232) and (9.218) by Roy and Smith (1969):

$$E[T_\theta] = \sum_{k=0}^\infty \frac{2^k}{(2k+1)!!(k+1)}(Y^{2k+2} - Z^{2k+2})$$
$$+ 2\sqrt{\pi}\left[Z\Phi\left(\frac{1}{2}, \frac{3}{2}; Z^2\right) - Y\Phi\left(\frac{1}{2}, \frac{3}{2}; Y^2\right)\right],$$

$$(9.238)$$

where

$$Y = (\mu - \theta)/\sigma, \quad (9.239)$$
$$Z = \mu/\sigma, \quad (9.240)$$

and where $(2k+1)!! = (2k+1)(2k-1), \ldots, 3.1$. An expression for the second moment has been derived by Ricciardi and Sacerdote (1979). Approximate (Thomas 1975), asymptotic (Sato 1978), and perturbation results (Wan and Tuckwell 1982) have been obtained for the mean and higher-order moments of the firing time.

In the last mentioned reference it is shown that, with a threshold normalized to unity, when the steady-state mean depolarization μ is much greater than threshold and

$$dx = (-x + \mu)\, dt + \epsilon\, dW, \quad (9.240A)$$

where $\epsilon \ll 1$, the approximate mean and variance of the firing time are

$$E[T] \simeq \ln\left(\frac{\mu}{\mu-1}\right) - \frac{\epsilon^2}{4}\left[\frac{1}{(\mu-1)^2} - \frac{1}{\mu^2}\right], \qquad (9.240B)$$

$$\text{Var}[T] \simeq \frac{\epsilon^2}{2}\left[\frac{1}{(\mu-1)^2} - \frac{1}{\mu^2}\right]. \qquad (9.240C)$$

The first of these formulas (9.240B) shows that a zero mean additive noise reduces the expected time at which the voltage reaches threshold.

Time-dependent thresholds have been employed by Geisler and Goldberg (1966), Matsuyama et al. (1974), Tuckwell (1981b), and Tuckwell and Wan (1984). For the special case of a threshold, which decays exponentially with the same time constant as the membrane's, the OUP barrier may be transformed to a Wiener process constant barrier, and an exact expression obtained for the firing-time density (Sugiyama et al. 1970; Clay and Goel 1973).

A comparison has been made of the expected ISI in Stein's model and for the OUP by Tuckwell and Cope (1980). The values of μ and σ were determined by (9.201) and (9.202). Some of these results are shown in Table 9.1. It is found that the diffusion approximation may sometimes severely underestimate and sometimes severely overestimate the mean ISI for the discontinuous process.

In Section 9.6 we showed that in the limit of small steps and large frequencies, the randomized random walk at time t, converged in distribution to $W(t)$. Much more than this can be shown. In fact, the process V_a can be shown to *converge weakly* (Billingsley 1968) to W as $a \to 0$. This implies that the *finite-dimensional distributions* of V_a converge to those of W. The weak convergence of the depolarization in Stein's model as steps go to zero and frequencies approach ∞ to the OUP has been established by Kallianpur (1983) Kallianpur and Wolpert (1984b), and Lansky (1984). According to Kurtz (1981) this weak convergence follows from the uniform convergence of the corresponding sequence of infinitesimal generators.

Diffusion approximations to discontinuous processes are expected to be most useful when the discontinuities are very small and very frequent. However, if $\hat{X}$ is a diffusion approximating X, error estimates for quantities such as mean first-passage times of $\hat{X}$ relative to those of X must be provided as well as the establishment of the weak convergence to $\hat{X}$.

An alternative approach is that of Walsh (1981a) who shows how a diffusion approximation may sometimes be constructed that has the

Table 9.1. *A comparison of the mean interspike intervals calculated for Stein's model (M_1) and the OUP diffusion approximation (M_1^*)*

θ	λ_I	λ_E	$M_1(0)$	$M_1^*(0)$
4	2	2	55.1	56.7
		3	10.4	9.39
		4	4.21	3.69
		5	2.40	2.10
4	6	3	324	195
		4	52.3	38.5
		5	15.7	12.5
		6	6.82	5.69
		7	3.77	3.21
		8	2.43	2.09
8	2	4	167	327
		5	33.0	40.6
		6	11.7	11.9
		7	5.92	5.60
		8	3.71	3.43
		9	2.64	2.42
		10	2.03	1.86
8	10	8	261	218
		9	81.7	70.4
		10	32.8	28.8
		11	16.0	14.2

Diffusion results are from formula (9.238). Results for the discontinuous process are from Cope and Tuckwell (1979).

same expected first-passage times as the original discontinuous process.

9.10   Stochastic cable theory

The stochastic models considered thus far in this chapter are analogous to the Lapicque model of Chapter 3 – they ignore the spatial extent of the nerve cell. Hence they cannot, for example, distinguish the effects of synaptic or other inputs that occur on different parts of the neuron. A stochastic version of nonlinear systems of equations such as those of Hodgkin and Huxley is a mathematically formidable task [but see Skaugen (1978) for simulation studies]. We have found that considerable progress can be made with stochastic versions of the linear cable model of Chapters 4–6. We begin with some generalities and then concentrate on some specific cases that have been considered recently.

9.10.1   A general stochastic cable equation: impulsive current inputs

We consider a nerve cylinder (possibly an equivalent cylinder representing a dendritic tree) on $[a, b]$ in which the potential satisfies the cable equation (cf. Chapter 4)

$$V_t = -V + V_{xx} + I, \qquad a < x < b, \, t > 0, \tag{9.241}$$

where $I = I(x, t)$ is the current density. [Note that we are working with dimensionless space and time and have set $\bar{c}_m = 1$ in Equation (4.56).] Now $I(x, t) \, dx \, dt$ is the charge passing through the element of length $(x, x + dx)$ in $(t, t + dt)$, and we recall from Chapter 4 that if $Q(x, t)$ is the charge through $(0, x)$ in $(0, t)$, then $I = Q_{xt}$. We extend the concept of Poisson random measure utilized in stochastic point models to include spatial dependence. We set

$$\frac{\partial^2 Q}{\partial x \, \partial t} \, dx \, dt = \int_{\mathbb{R}} u\nu(dt, dx, du) = I(x, t) \, dx \, dt. \tag{9.242}$$

We assume ν has the following properties.

(i) Let $A \subset [a, b]$, $B \subset \mathbb{R}$. Then $\{\nu(t, A, B), \, t \geq 0\}$ is a temporally homogeneous Poisson process with mean rate $\Pi(A, B)$.

(ii) Let A_i, $i = 1, \ldots, m$, and B_j, $j = 1, \ldots, n$, be disjoint subsets of $[a, b]$ and $\mathbb{R}$, respectively. Then the collection of random variables $\nu(t, A_i, B_j)$, $i = 1, \ldots, m$; $j = 1, \ldots, n$, are mutually independent.

To see what this means, set

$$Y(t, x) = \int_{\mathbb{R}} u\nu(t, x, du) = \int_0^t \int_a^x \int_{\mathbb{R}} u\nu(dt', dx', du), \tag{9.243}$$

which is a compound Poisson process in two dimensions. Note that

$$\nu(t, A, B) = \int_A \int_B \nu(t, dx, du) \tag{9.244}$$

counts the number of jumps of Y up to time t that occur in the space subset A and have amplitudes in B. The total mean rate of arrival of jumps is

$$\Lambda = \int_a^b \int_{\mathbb{R}} \Pi(dx, du). \tag{9.245}$$

We will usually assume that Π has a density so that we may formally write $\Pi(dx, du) = \phi(x, u) \, dx \, du$

Solution of the cable equation and the moments of the depolarization

Suppose there are given boundary conditions at $x = a$ and $x = b$ and that the initial depolarization is

$$V(x, 0) = V_0(x), \qquad a \leq x \leq b, \tag{9.246}$$

possibly random. Assuming the boundary conditions are such that the

Green's function method of solution (cf. Section 5.2) may be employed, then

$$V(x,t) = \int_0^L G(x,y;t)V_0(y)\,dy$$
$$+ \int_0^t \int_a^b \int_{-\infty}^\infty G(x,y;t-s)u\nu(ds,dy,du). \quad (9.247)$$

For convenience in what follows we let the initial condition be that of a resting cell, so $V_0(x) = 0$, $x \in [a,b]$. The expectation of $V(x,t)$ is

$$E[V(x,t)] = \int_0^t \int_a^b \int_{-\infty}^\infty G(x,y;t-s)u\phi(y,u)\,dy\,du\,ds, \quad (9.248)$$

and its variance is

$$\text{Var}[V(x,t)] = \int_0^t \int_a^b \int_{-\infty}^\infty G^2(x,y;t-s)u^2\phi(y,u)\,dy\,du\,ds. \quad (9.249)$$

The total mean rate of arrival of impulses in $(x, x+dx)$ is

$$\lambda(x)\,dx = \int_{-\infty}^\infty \phi(x,u)\,du\,dx, \quad (9.250)$$

and the average amplitude in $(x, x+dx)$ is

$$\mu(x) = \frac{\int_{-\infty}^\infty u\phi(x,u)\,du}{\lambda(x)}. \quad (9.251)$$

Furthermore, the variance of the amplitude is

$$\sigma^2(x) = \frac{\int u^2\phi(x,u)\,du}{\lambda(x)} - \mu^2(x). \quad (9.252)$$

In terms of these quantities,

$$E[V(x,t)] = \int_0^t \int_a^b G(x,y;t-s)\mu(y)\lambda(y)\,dy\,ds, \quad (9.253)$$

$$\text{Var}[V(x,t)] = \int_0^t \int_a^b G^2(x,y;t-s)[\mu^2(y)+\sigma^2(y)]\lambda(y)\,dy\,ds. \quad (9.254)$$

Simple Poisson excitation and inhibition

Let the nerve cylinder receive Poisson excitation at x_E with mean rate λ_E and amplitude a_E and Poisson inhibition at x_I with mean rate λ_I and amplitude a_I. The rate density is then

$$\phi(x,u) = \lambda_E\delta(x-x_E)\delta(u-a_E) + \lambda_I\delta(x-x_I)\delta(u+a_I). \quad (9.255)$$

Substitution of ϕ in the above formulas gives

$$E[V(x,t)] = a_E \lambda_E \int_0^t G(x, x_E; t-s)\, ds$$

$$- a_I \lambda_I \int_0^t G(x, x_I; t-s)\, ds, \qquad (9.256)$$

$$\mathrm{Var}[V(x,t)] = a_E^2 \lambda_E \int_0^t G^2(x, x_E; t-s)\, ds$$

$$+ a_I^2 \lambda_I \int_0^t G^2(x, x_I; t-s)\, ds. \qquad (9.257)$$

Note that in terms of the simple Poisson processes N_E and N_I the stochastic cable equation can be written as

$$\frac{\partial V}{\partial t} = -V + \frac{\partial^2 V}{\partial x^2} + a_E \delta(x - x_E) \frac{dN_E}{dt} - a_I \delta(x - x_I) \frac{dN_I}{dt}. \qquad (9.258)$$

In the case of multiple excitatory and inhibitory input sites, we have

$$\frac{\partial V}{\partial t} = -V + \frac{\partial^2 V}{\partial x^2} + \sum_{i=1}^m a_{E,i} \delta(x - x_{E,i}) \frac{dN_{E,i}}{dt}$$

$$- \sum_{j=1}^n a_{I,j} \delta(x - x_{I,j}) \frac{dN_{I,j}}{dt}, \qquad (9.259)$$

and the mean and variance are given by (9.256) and (9.257) with $a_E \lambda_E$ replaced by $\sum_{i=1}^m a_{E,i} \lambda_{E,i}$ and $a_E^2 \lambda_E$ replaced by $\sum_{i=1}^m a_{E,i}^2 \lambda_{E,i}$; similarly for the inhibitory contribution.

Mean and variance for various boundary conditions

We utilize the expressions obtained in Chapters 5 and 6 for the cable equation Green's functions. Only a few cases are considered.

(i) *Infinite cylinder*, $x \in (-\infty, \infty)$.   The Green's function is given as case (i) in Section 5.6. Accordingly, the mean depolarization is, on using (5.48),

$$E[V(x,t)] = \frac{1}{2\sqrt{2}} \left[a_E \lambda_E \left\{ e^{-|x-x_E|} \mathrm{erfc}\left(\frac{|x-x_E| - 2t}{2\sqrt{t}} \right) \right. \right.$$

$$\left. - e^{|x-x_E|} \mathrm{erfc}\left(\frac{|x-x_E| + 2t}{2\sqrt{t}} \right) \right\}$$

$$- a_I \lambda_I \left\{ e^{-|x-x_I|} \mathrm{erfc}\left(\frac{|x-x_I| - 2t}{2\sqrt{t}} \right) \right.$$

$$\left. \left. - e^{|x-x_I|} \mathrm{erfc}\left(\frac{|x-x_I| + 2t}{2\sqrt{t}} \right) \right\} \right]. \qquad (9.260)$$

This is the deterministic result with a constant current at x_E and x_I. As $t \to \infty$, the steady-state value,

$$E[V(x, \infty)] = \frac{1}{\sqrt{2}} \left[a_E \lambda_E e^{-|x-x_E|} - a_I \lambda_I e^{-|x-x_I|} \right] \quad (9.261)$$

is approached.

An explicit expression for the variance at time t seems difficult to obtain. The following formula holds:

$$\text{Var}[V(x, t)] = \frac{a_E^2 \lambda_E}{4\pi} \int_0^t \frac{e^{-2T} e^{-(x-x_E)^2/2T}}{T} dT$$

$$+ \frac{a_I^2 \lambda_I}{4\pi} \int_0^t \frac{e^{-2T} e^{-(x-x_I)^2/2T}}{T} dT. \quad (9.262)$$

Using a standard integral (Gradshteyn and Ryzhik 1965, page 340), the steady-state variance is found to be

$$\text{Var}[V(x, \infty)] = \frac{1}{2\pi} \left[a_E^2 \lambda_E K_0(2|x - x_E|) + a_I^2 \lambda_I K_0(2|x - x_I|) \right],$$

$$(9.263)$$

where $K_0(z)$ is a modified Bessel function (Abramowitz and Stegun 1965, page 374). Note that $K_0(x)$ is singular at the origin so that the variance is unbounded at $x = x_E$ and $x = x_I$.

(ii) *Semiinfinite cylinders*, $x \in [0, \infty)$. The Green's functions for semiinfinite cylinders with sealed and killed ends at $x = 0$ are given by (5.67) and (5.68). These are sums and differences of the infinite cable Green's function. Thus, for example, with a sealed end at $x = 0$ and a single excitatory input at $x = x_E$,

$$E[V(x, t)] = \frac{1}{2\sqrt{2}} a_E \lambda_E \left\{ e^{-|x-x_E|} \text{erfc}\left(\frac{|x - x_E| - 2t}{2\sqrt{t}} \right) \right.$$

$$- e^{|x-x_E|} \text{erfc}\left(\frac{|x - x_E| + 2t}{2\sqrt{t}} \right)$$

$$+ e^{-|x+x_E|} \text{erfc}\left(\frac{|x + x_E| - 2t}{2\sqrt{t}} \right)$$

$$\left. - e^{|x+x_E|} \text{erfc}\left(\frac{|x + x_E| + 2t}{2\sqrt{t}} \right) \right\}, \quad (9.264)$$

which as $t \to \infty$ approaches

$$E[V(x, \infty)] = \frac{1}{\sqrt{2}} a_E \lambda_E \left(e^{-|x-x_E|} + e^{-|x+x_E|} \right). \quad (9.265)$$

The asymptotic, $t \to \infty$, variance is

$$\text{Var}[V(x,\infty)] = \frac{a_E^2 \lambda_E}{2\pi} \Big[K_0(2|x - x_E|) + K_0(2|x + x_E|)$$

$$+ 2K_0\big(2\sqrt{x^2 + x_E^2}\big)\Big].$$

$$(9.266)$$

For the semiinfinite cylinder with a lumped soma attached at $x = 0$ the Green's function involves the complementary error function and no closed-form expressions seem available for the mean and variance.

(iii) *Finite cables*, $x \in [0, L]$. We have seen in Chapter 5 that there are two equivalent expressions for the Green's function for the finite cable with boundary conditions that are combinations of the sealed- and killed-end conditions. One representation was obtained by the method of images; the other was an eigenfunction expansion. When the former is used, the mean and variance of the depolarization are expressible as a sum of terms like (9.260). The eigenfunction expansions lead to infinite and doubly infinite series for the mean and variance as the following examples show. We only consider the case of a single excitatory input at x_E.

Consider a nerve cylinder with sealed ends at $x = 0$ and $x = L$. The following results can be obtained from those for white-noise current injection at $x = x_E$ (Wan and Tuckwell 1979):

$$E[V(x,t)] = a_E \lambda_E \sum_{n=0}^{\infty} \frac{1 - e^{-k_n^2 t}}{\mu_n^2} \phi_n(x_E)\phi_n(x), \quad (9.267)$$

$$E[V(x,\infty)] = \begin{cases} a_E \lambda_E \dfrac{\cosh(L - x_E)\cosh x}{\sinh L}, & x < x_E, \\[2ex] a_E \lambda_E \dfrac{\cosh x_E \cosh(L - x)}{\sinh L}, & x > x_E, \end{cases}$$

$$(9.268)$$

$$\text{Var}[V(x,\infty)] = a_E^2 \lambda_E \sum_{m=0}^{\infty} \sum_{n=0}^{\infty} \frac{\phi_m(x_E)\phi_n(x_E)\phi_m(x)\phi_n(x)}{k_m^2 + k_n^2},$$

$$(9.269)$$

where $\{\phi_n, \ n = 0,1,\ldots\}$ are the spatial eigenfunctions (5.38) and k_n^2 are the corresponding eigenvalues (5.27) and (5.31). The variance is infinite at the point of application of the stimulus.

Figure 9.23 shows the temporal evolution of the mean and variance for Poisson excitation of a cable of length $L = 1$ with $a_E = \lambda_E = 1$ for

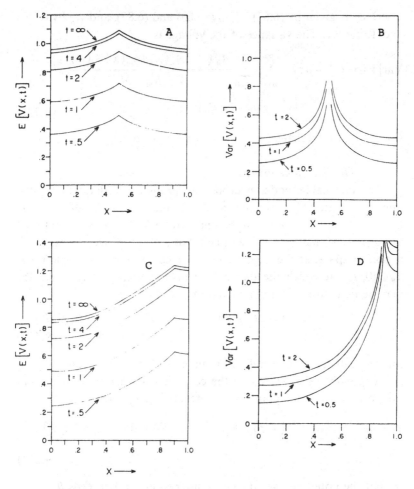

Figure 9.23. Mean and variance of the depolarization at various times for a nerve cylinder undergoing Poisson stimulation. In all cases $a_E = \lambda_E = L = 1$. A – Mean with $x_E = 0.5$. B – Variance with $x_E = 0.5$. C – Mean with $x_E = 0.9$. D – Variance with $x_E = 0.9$. [From Wan and Tuckwell (1979). Reproduced with the permission of Springer-Verlag and the authors.]

two different input positions. One of these is at the center of the cylinder, the other being remote from the origin at $x = 0.9$.

Consider now a nerve cylinder of length L with a lumped soma affixed at $x = 0$ and a sealed end at $x = L$. The Green's function is given by (6.86). Substituting in (9.256) gives

$$E[V(x,t)] = a_E \lambda_E \sum_{n=0}^{\infty} \frac{A_n(x_E)\phi_n(x)}{1+\lambda_n^2} \left[1 - e^{-t(1+\lambda_n^2)} \right], \qquad x < x_E,$$

(9.270)

where ϕ_n is given by (6.67), A_n by (6.83) and (6.85), and λ_n by (6.60) and Table 6.1. The variance of the voltage is

$$\text{Var}[V(x,t)] = a_E^2 \lambda_E \sum_{m=0}^{\infty} \sum_{n=0}^{\infty} \frac{A_m(x_E)A_n(x_E)\phi_m(x)\phi_n(x)}{2 + \lambda_m^2 + \lambda_n^2}$$

$$\times \left[1 - e^{-t(2+\lambda_m^2+\lambda_n^2)}\right], \qquad x < x_E. \qquad (9.271)$$

For $x \geq x_E$, interchange x and x_E in these formulas.

The interspike interval

For cable-model neurons the problem of determining the interspike interval is complicated by the multitude of possible threshold conditions that can be imposed on V. For a nerve on $[0, L]$ a trigger zone may be assumed to exist at a point or over an extended region in space. If the trigger zone is taken to be at the single point $x_\theta \in [0, L]$, at which the threshold depolarization is θ, then the ISI may be defined as the random variable

$$T_\theta = \inf\{t \,|\, V(x_\theta, t) \geq \theta\}, \qquad V(x,0) = 0, \qquad x \in [0, L], \qquad (9.272)$$

for an initially resting cell. For an extended trigger zone, say in $[x_1, x_2]$, with $0 \leq x_1 < x_2 \leq L$, the condition for firing could be that the potential at some point in the interval $[x_1, x_2]$ exceeds θ,

$$T_\theta = \inf\left\{ t \,\middle|\, \sup_{x_1 \leq x \leq x_2} V(x,t) \geq \theta \right\}, \qquad V(x,0) = 0, \qquad x \in [0, L], \qquad (9.273)$$

or that the potential over the entire interval $[x_1, x_2]$ exceeds θ,

$$T_\theta = \inf\left\{ t \,\middle|\, \inf_{x_1 \leq x \leq x_2} V(x,t) \geq \theta \right\}, \qquad V(x,0) = 0, \qquad x \in [0, L]. \qquad (9.274)$$

These three kinds of threshold conditions are illustrated in Figure 9.24. One could also devise threshold conditions on the axial current in terms of $\partial V/\partial x$ or even in terms of both V and $\partial V/\partial x$. There is a perplexing variety of possible conditions, and exact solutions for the moments and distribution of T_θ are difficult if not impossible to obtain. Simulation, it seems, is a very powerful tool compared to exact or analytic techniques in this kind of problem. The work that has been performed thus far has assumed the simplest threshold condition (9.272) with $x_\theta = 0$ (Tuckwell and Walsh 1983; Tuckwell, Wan, and Wong 1984).

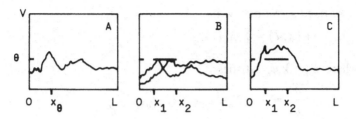

Figure 9.24. Three kinds of threshold conditions for the cable-model neuron. A, B, and C correspond to Equations (9.272), (9.273), and (9.274), respectively.

Infinite series of discontinuous Markov processes

We consider briefly the case of a nerve cylinder on $[0, L]$ with Poisson excitation at x_E. Then

$$\frac{\partial V}{\partial t} = -V + \frac{\partial^2 V}{\partial x^2} + a_E \delta(x - x_E)\frac{dN_E}{dt}, \qquad 0 < x < L, \, t > 0.$$
(9.275)

We will show that at a given space point the voltage can be represented as an infinite series of discontinuous Markov processes. Each component in the sum is analogous to the depolarization in Stein's model. First, we assume that the Green's function can be expressed as

$$G(x, y; t) = \sum_n \phi_n(x)\phi_n(y)e^{-\lambda_n t}.$$
(9.276)

The solution of (9.275) is thus

$$V(x, t) = a_E \sum_n \phi_n(x)\phi_n(x_E)e^{-\lambda_n t}\int_0^t e^{\lambda_n s}\frac{dN_E}{ds}\,ds.$$
(9.277)

Now define

$$V_n(t) = a_E \phi_n(x_E)e^{-\lambda_n t}\int_0^t e^{\lambda_n s}\frac{dN_E}{ds}\,ds,$$
(9.278)

so that

$$V(x, t) = \sum_n V_n(t)\phi_n(x).$$
(9.279)

Then V_n is seen to be the solution of the ordinary stochastic differential equation

$$\frac{dV_n}{dt} = -\lambda_n V_n + a_E \phi_n(x_E)\frac{dN_E}{dt},$$
(9.280)

which should be compared with (9.140). Thus V can be represented as an infinite series of discontinuous Markov processes.

At fixed x, we may write

$$V(x, t) = \sum_n X_n(t),$$ (9.281)

where $X_n = V_n \phi_n$ satisfies

$$\frac{dX_n}{dt} = -\lambda_n X_n + a_E \phi_n(x_E) \phi_n(x) \frac{dN_E}{dt}.$$ (9.282)

We may approximate the infinite series (9.281) by retaining a finite number of terms. If the first m terms are retained, then the *approximate ISI* is

$$T_{\theta, m} = \inf\left\{ t \,\middle|\, \sum_{n=1}^{m} X_n(t) \geq \theta \right\}.$$ (9.283)

The theory of vector-valued Markov processes may thus be applied to determine the moments and distribution function of $T_{\theta, m}$. We illustrate with $m = 2$.

Put $\alpha_n = a_E \phi_n(x_E) \phi_n(x)$ and let $\mathbf{X} = (X_1, X_2)$. The infinitesimal generator of $\mathbf{X}$ is given by

$$(\mathscr{A}f)(x_1, x_2) = \lim_{\Delta t \downarrow 0} \frac{E[f(\mathbf{X}(t + \Delta t)) - f(\mathbf{X}(t)) | \mathbf{X}(t) = x]}{\Delta t}.$$ (9.284)

A calculation from first principles gives

$$(\mathscr{A}f)(x_1, x_2) = -\lambda_1 x_1 \frac{\partial f(x_1, x_2)}{\partial x_1} - \lambda_2 x_2 \frac{\partial f(x_1, x_2)}{\partial x_2}$$
$$+ \lambda_E [f(x_1 + \alpha_1, x_2 + \alpha_2) - f(x_1, x_2)].$$ (9.285)

Now let $\mathbf{X}(0) = (x_1, x_2)$ and let

$$T_{\theta, 2}(x_1, x_2) = \inf\{t | X_1(t) + X_2(t) \geq \theta\}, \qquad x_1 + x_2 < \theta.$$ (9.286)

Then, if the nth moment of $T_{\theta, 2}(x_1, x_2)$ is denoted by $\mu_n(x_1, x_2)$, we have the recursion system

$$-\lambda_1 x_1 \frac{\partial \mu_n}{\partial x_1} - \lambda_2 x_2 \frac{\partial \mu_n}{\partial x_2} + \lambda_E [\mu_n(x_1 + \alpha_1, x_2 + \alpha_2) - \mu_n(x_1, x_2)]$$

$$= -n\mu_{n-1}, \qquad n = 0, 1, 2 \ldots, \quad x_1 + x_2 < \theta.$$ (9.287)

For the first- and higher-order moments the boundary condition is

$$\mu_n(x_1, x_2) = 0, \qquad x_1 + x_2 \geq \theta, \qquad n = 1, 2, \ldots.$$ (9.288)

The moments of the firing time for a cell initially at rest are approximated by $\mu_n(0, 0)$. Similarly, one may treat the case of excitation and inhibition and of random PSP amplitudes.

If the boundary condition at $x = 0$ is that of a *lumped soma*, so that, for example, with $x < y$,

$$G(x, y; t) = \sum_n A_n(y)\phi_n(x)e^{-\lambda_n t}, \tag{9.289}$$

then the only modification is that (9.282) is replaced by

$$\frac{dX_n}{dt} = -\lambda_n X_n + a_E A_n(x_E)\phi_n(x)\frac{dN_E}{dt}, \qquad x < x_E. \tag{9.290}$$

Uniform Poisson stimulation

If the rate of arrival of impulses is the same in all elements of length along the cable, then the rate density is

$$\phi(x, u) = \lambda_E \delta(u - a_E) + \lambda_I \delta(u + a_I), \tag{9.291}$$

independent of x. The total mean rate of arrival of impulses on the whole cable is

$$\Lambda = \int_a^b \int_{\mathbf{R}} \phi(x, u)\, dx\, du = (b - a)(\lambda_E + \lambda_I). \tag{9.292}$$

The cable equation may be written as

$$\frac{\partial V}{\partial t} = -V + \frac{\partial^2 V}{\partial x^2} + a_E \frac{\partial^2 N_E}{\partial x\, \partial t} - a_I \frac{\partial^2 N_I}{\partial x\, \partial t}, \qquad a < x < b,\ t > 0, \tag{9.293}$$

where $N_E(x, t)$ and $N_I(x, t)$ are two independent Poisson processes in the strip $[a, b] \times [0, \infty)$.

From (9.248) and (9.249) the mean and variance of the depolarization are

$$E[V(x, t)] = (\lambda_E a_E - \lambda_I a_I)\int_0^t \int_a^b G(x, y; t - s)\, dy\, ds, \tag{9.294}$$

$$\mathrm{Var}[V(x, t)] = (\lambda_E a_E^2 + \lambda_I a_I^2)\int_0^t \int_a^b G^2(x, y; t - s)\, dy\, ds. \tag{9.295}$$

Many of the results for uniform Poisson stimulation are the same as for uniform white noise, which is considered in the next section.

9.10.2   Stochastic cable equation – white-noise currents

(A) *White-noise current injection at a point*

With white-noise current applied at $x = x_0$, we may write

$$\frac{\partial V}{\partial t} = -V + \frac{\partial^2 V}{\partial x^2} + \delta(x - x_0)\left[\alpha + \beta\frac{dW}{dt}\right], \qquad a < x < b,\ t > 0, \tag{9.296}$$

where α and β are constants and W is a standard Wiener process.

Note that the "derivative" of W, called white noise, in the equation implies a forthcoming integration. Equation (9.296) has been studied by Wan and Tuckwell (1979) and Tuckwell et al. (1984). The case of white-noise current injection over a small length of the cable was treated in Tuckwell and Wan (1980).

The solution of (9.296) under the assumption of zero initial depolarization everywhere is

$$V(x,t) = \int_0^t \int_a^b G(x,y;t-s)\left[\alpha + \beta \frac{dW(s)}{ds}\right]\delta(y-x_0)\,ds\,dy$$

$$= \alpha \int_0^t G(x,x_0;t-s)\,ds + \beta \int_0^t G(x,y;t-s)\,dW(s).$$

$$(9.297)$$

The mean and variance of V satisfying (9.296) are the same as for Poisson stimulation with the replacements $a_E\lambda_E \to \alpha$ and $a_E\sqrt{\lambda_E} \to \beta$ in, for example, (9.268) and (9.269). The *covariance* of $V(x,t)$ and $V(y,\tau)$ may be found on utilizing the formal covariance property of white noise:

$$\mathrm{Cov}\left[\frac{dW(s)}{ds}, \frac{dW(t)}{dt}\right] = \delta(t-s). \qquad (9.298)$$

It is left as an exercise to show that the required covariance is

$$K(x,t;\,y,\tau) = \beta^2 \int_0^\tau G(x,x_0;t-s)G(y,x_0;\tau-s)\,ds. \qquad (9.299)$$

This result also holds for Poisson input at a point with $\beta^2 = a_E^2\lambda_E$.

Just as the solution of the cable equation on a finite interval with Poisson input at a point can be decomposed into an infinite number of discontinuous Markov processes, when the stimulus is white noise, the voltage can be expressed as an infinite series whose terms are Ornstein–Uhlenbeck processes. At fixed x,

$$V(x,t) = \sum_n X_n(t), \qquad (9.300)$$

where the X_n's satisfy

$$dX_n = (-\lambda_n X_n + \alpha\phi_n(x_0)\phi_n(x))\,dt + \beta\phi_n(x_0)\phi_n(x)\,dW. \qquad (9.301)$$

That is, each term in (9.300) is an Ornstein–Uhlenbeck process.

The time to nerve-cell firing has been investigated for (9.296) with a threshold condition of type (9.272) (Tuckwell et al. 1984). Numerical calculations were performed on the equations for the moments of the first-exit time using the first two terms in (9.300). These calculations

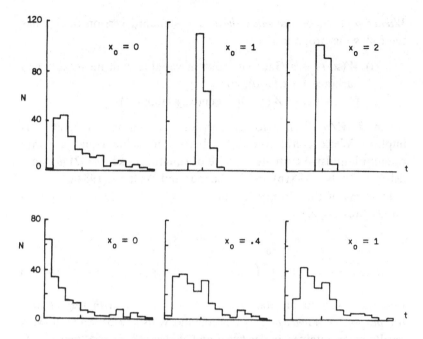

Figure 9.25. ISI histograms obtained by simulation of the depolarization of the solution of the cable equation with white-noise current at x_0. The bin widths are one-fifth of the mean interval, the latter being marked on the time axis. For the upper set of results, $\alpha = 10$, $\beta = 1$, $L = 2$, and $\theta = \sqrt{2}$; for the lower set, $\alpha = 20$, $\beta = 10$, $L = 1$, and $\theta = 10$. [From Tuckwell et al. (1984). Reproduced with the permission of Springer-Verlag and the authors.]

were complemented by simulation of the Ornstein–Uhlenbeck processes and by direct simulation of the stochastic integral solution (9.297). The simulations also enabled the ISI density to be estimated and some of the histograms obtained are shown in Figure 9.25. Notice the trend from exponential-like to gammalike, to normal-like, as the input location becomes more remote from the trigger zone. In these studies it was also found that for fixed values of the remaining parameters, the coefficient of variation of the ISI is a monotonically decreasing function of distance between source and trigger zone.

(B) *Two-parameter white noise*
A continuous version of (9.293) is

$$\frac{\partial V}{\partial t} = -V + \frac{\partial^2 V}{\partial x^2} + \alpha + \beta \frac{\partial^2 W}{\partial x \, \partial t}, \qquad a < x < b, \, t > 0,$$

$$(9.302)$$

where α and β are constants and $W = \{W(x,t)\}$ is a *two-parameter*

Wiener process, or *Brownian sheet*. The standard version of this has the following properties:

(i) $W(x, t)$ is a Gaussian random variable with mean zero and variance $xt(x, t \geq 0)$; and

(ii) $\text{Cov}[W(x, s), W(y, t)] = \min(x, y)\min(s, t)$.

Again $\partial^2 W / \partial x\, \partial t$ is a formal derivative – an ensuing integration is implied. Weak convergence of solutions to cable equations with random impulsive currents to those of equations like (9.302) is established in Walsh (1981b) and Kallianpur and Wolpert (1984a, b).

In terms of the Green's function the solution of (9.302) is, for an initially resting cell,

$$V(x, t) = \alpha \int_a^b \int_0^t G(x, y; t - s)\, ds\, dy$$

$$+ \beta \int_a^b \int_0^t G(x, y; t - s)\, dW(s, y), \qquad (9.303)$$

where the second integral is a *stochastic integral with respect to a two-parameter Wiener process* (Cairoli and Walsh 1975). The following results were obtained in Tuckwell and Walsh (1983) and their proofs are left as exercises for the reader.

(i) *An infinite cable.* The *mean* and *variance* of the depolarization are

$$E[V(x, t)] = \alpha(1 - e^{-t}) \underset{t \to \infty}{\sim} \alpha, \qquad (9.304)$$

$$\text{Var}[V(x, t)] = \frac{\beta^2}{4}\left[1 - \text{erfc}(\sqrt{2t})\right] \underset{t \to \infty}{\sim} \frac{\beta^2}{4}. \qquad (9.305)$$

The covariance of $V(x, t)$ and $V(y, \tau)$ is, with $\tau \geq t$,

$$K(x, t; y, \tau) = \frac{\beta^2}{8}\Bigg[e^{-|x-y|}\bigg\{ \text{erfc}\left(\frac{|x - y| - 2(t + \tau)}{2\sqrt{t + \tau}} \right) $$

$$- \text{erfc}\left(\frac{|x - y| - 2(\tau - t)}{2\sqrt{\tau - t}} \right) \bigg\} $$

$$- e^{|x-y|}\bigg\{ \text{erfc}\left(\frac{|x - y| + 2(t + \tau)}{2\sqrt{t + \tau}} \right) $$

$$- \text{erfc}\left(\frac{|x - y| + 2(\tau - t)}{2\sqrt{\tau - t}} \right) \bigg\}\Bigg]. \qquad (9.306)$$

If we evaluate $K(x, t; x, t + \tau)$ and let $\tau \to \infty$,

$$K(x, t; x, t + \tau) \underset{\tau \to \infty}{\longrightarrow} \frac{\beta^2}{4}\text{erfc}(\sqrt{\tau}) \doteq \bar{K}(\tau). \qquad (9.307)$$

Thus at fixed x the depolarization becomes asymptotically *wide-sense stationary* (or *covariance stationary*). We may therefore compute the *spectral density* of the long-term depolarization

$$f(\omega) = \frac{1}{2\pi} \int_{-\infty}^{\infty} \exp[-i\omega\tau] \overline{K}(\tau)\, d\tau. \qquad (9.308)$$

Evaluating the integrals gives

$$f(\omega) = \frac{\beta^2}{4\sqrt{2}\,\pi} \frac{1}{\omega} \cdot \frac{\sqrt{(1+\omega^2)^{1/2} - 1}}{(1+\omega^2)^{1/2}} \underset{\omega \to \infty}{\sim} \frac{\beta^2\sqrt{2}\,\omega^{-3/2}}{8\pi}. \qquad (9.309)$$

Hence at high frequencies the spectrum becomes proportional to the frequency raised to the power $-3/2$ ("1 over $f^{3/2}$").

(ii) *Finite cables.* For finite cables the solution of (9.302) can be written as an infinite series in which each term is an Ornstein–Uhlenbeck process. In this case, however, and in distinction to the case of white-noise injection at a point, each OUP is independent of all the others. Assuming a cable on $[0, L]$, we find the *mean voltage* is

$$E[V(x, t)] = \alpha(1 - e^{-t}), \qquad (9.310)$$

for *sealed ends*, whereas it is

$$E[V(x, t)] = \frac{4\alpha}{\pi} \sum_{n=1}^{\infty}{}' \frac{\sin(n\pi x/L)}{n[1 + n^2\pi^2/L^2]} \left\{ 1 - \exp\left[-\left(1 + \frac{n^2\pi^2}{L^2}\right)t \right] \right\}, \qquad (9.311)$$

where $\sum'$ denotes summation over odd n only, for *killed ends*. In general, the variance is

$$\mathrm{Var}[V(x, t)] = \beta^2 \sum_{n} \frac{\phi_n^2(x)}{2\lambda_n} [1 - \exp\{-2\lambda_n t\}], \qquad (9.312)$$

where ϕ_n and λ_n are the eigenfunctions and corresponding eigenvalues. For *sealed ends* we find

$$\mathrm{Var}[V(x, \infty)] = \beta^2 \frac{\cosh(L - x)\cosh x}{2\sinh L}, \qquad (9.313)$$

whereas for *killed ends*

$$\mathrm{Var}[V(x, \infty)] = \beta^2 \frac{\sinh(L - x)\sinh x}{2\sinh L}. \qquad (9.314)$$

The steady-state mean and variance are sketched in Figure 9.26.

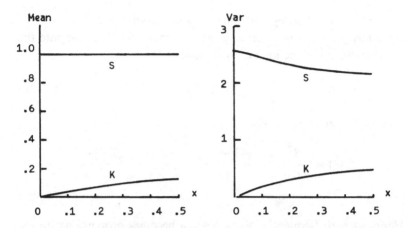

Figure 9.26. The steady-state mean and variance of the depolarization for a finite cable with uniform white-noise stimulation in the case of sealed (S) ends and killed (K) ends. $L = \alpha = \beta = 1$. [From Tuckwell and Walsh (1983). Reproduced with the permission of Springer-Verlag and the authors.]

The spectral density for the finite cable, corresponding to (9.309), can be obtained explicitly for some values of x. At the center of the cable,

$$
f(\omega)|_{x=L/2} = \frac{\beta^2}{4\pi\omega(1+\omega^2)^{1/2}}
$$

$$
\times \left[\frac{\gamma\left(\dfrac{\sin(2L\rho)}{2} \pm \cosh(L\gamma)\sin(L\rho)\right) + \rho\left(\dfrac{\sinh(2L\gamma)}{2} \pm \sinh(L\gamma)\cos(L\rho)\right)}{\sinh^2(L\gamma) + \sin^2(L\rho)} \right],
$$

$$(9.315)$$

where

$$
\gamma = \sqrt{\left[(1+\omega^2)^{1/2} + 1\right]/2}, \qquad \rho = \sqrt{\left[(1+\omega^2)^{1/2} - 1\right]/2},
$$

and where "$+$" refers to sealed ends and "$-$" refers to killed ends. Figure 9.27 shows the spectral densities for various L and for the cases of sealed and killed ends. As $L \to \infty$ the spectral densities

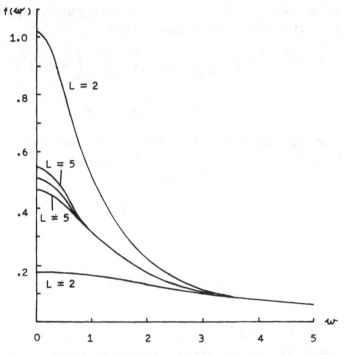

Figure 9.27. Spectral density functions for various cable lengths and boundary conditions for uniform white-noise current. Results for finite cables are for the center. The middle curve is for the infinite cable, those above being for sealed ends, those below for killed ends. [From Tuckwell and Walsh (1983). Reproduced with the permission of Springer-Verlag and the authors.]

approach that for the infinite cable, the approach being from below for killed ends and from above for sealed ends.

(C) *Neurons with dendritic trees*

Within the constraints imposed by the equivalent-cylinder concept (see Chapters 4 and 5), the results obtained for nerve cylinders may be applied to neurons with dendritic trees. The mapping procedure given in Section 5.10 may be used to determine the potential over an entire dendritic tree if the three-halves power law is obeyed at branch points. An exact expression has recently been obtained for the somatic depolarization in a nerve cell with several such dendritic trees when there is white-noise current injected at various locations over the neuron surface (Tuckwell 1984).

Finally, a full cable model for the stochastic activity of a nerve cell must include reversal potentials for excitatory and inhibitory synaptic action. Then on each dendritic segment the depolarization will satisfy an equation of the kind [cf. Equation (7.31)]

$$\frac{\partial V}{\partial t} = -V + \frac{\partial^2 V}{\partial x^2} + (V_E - V)\frac{\partial^2}{\partial x \partial t}\int_{\mathbf{R}} u\nu_E(t, x, du)$$

$$+ (V_I - V)\frac{\partial^2}{\partial x \partial t}\int_{\mathbf{R}} u\nu_I(t, x, du), \tag{9.316}$$

where $\nu_{E, I}$ are the Poisson random measures associated with excitation and inhibition. For stimulation at single excitatory and inhibitory sites, these equations will reduce to

$$\frac{\partial V}{\partial t} = -V + \frac{\partial^2 V}{\partial x^2} + (V_E - V)a_E\delta(x - x_E)\frac{dN_E}{dt}$$

$$+ (V_I - V)a_I\delta(x - x_I)\frac{dN_I}{dt}. \tag{9.317}$$

Such equations have not yet been the subject of any investigations.

Additional notes

There are a number of related topics not included in this chapter. The details of sequences of *miniature endplate potentials* have been reviewed by, for example, Van der Kloot et al. (1975). *Membrane noise*, which we briefly touched on in Section 9.10 and which includes *channel noise*, has been reviewed by Stevens (1972, 1977), Verveen and DeFelice (1974), DeFelice (1977), Neher and Stevens (1977), Holden (1981, 1983), and Neher (1983). *Continuous-time Markov chain* models for the random opening and closing of ion channels have been analyzed by Colquhoun and Hawkes (1981, 1982) and Jackson (1985). Hille (1984) contains a wonderful account of this subject. A diffusion approximation is presented in Tuckwell (1987a). *Wiener's expansion* for the response of a system to white noise is dealt with in Marmarelis and Marmarelis (1978). Many of these topics are the subject of a recent monograph Tuckwell 1988b). A number of *neural models* including *discrete* ones [see, for example, Goel, Richter-Dyn, and Clay (1972)], those whose *inputs last a random time* [see, for example, Ten Hoopen (1967)], and others (White and Ellias 1979) have not been considered in this study. Some are included in Holden (1976b) and Lee (1979). Nonlinear reaction–diffusion equations with additive white noise have been investigated both with one component (Tuckwell 1987b) and two components (Tuckwell 1987c). Surprisingly, similar equations have risen in quantum field theory.

10

The analysis of stochastic neuronal activity

10.1 Introduction

Given a record of the spiking activity of a neuron, the question arises as to what information can be obtained from it. There are many aspects to this question. A primary distinction must be made, of course, between *spontaneous* and *driven* activity.

It is a fact that the majority of CNS cells sampled thus far exhibit spontaneous activity. For example, in the cat cochlear nucleus, Pfeiffer and Kiang (1965) found that 80% of the 269 cells studied spontaneously emitted spikes. It is tacitly assumed in most studies that the spontaneous activity is not significantly affected by small doses of anaesthetic. This has been demonstrated for cells of Clarke's column (Pyatigorskii 1966).

The spontaneous, or background, activity is an important object to study for several reasons. First, it is a property of the cell and is of interest in itself. Furthermore, in conjunction with physiologically realistic mathematical models of nerve cells, it may be possible to infer actual nerve-cell properties from the spontaneous activity. Attempts of this kind have been made by Bishop et al. (1964), Braitenberg et al. (1965), Smith and Smith (1965), Correia and Landolt (1977), Tuckwell and Richter (1978), and Tuckwell (1979a). Statistical estimation procedures for parameters in some neuronal random processes have been outlined by Lansky (1983) and will be elaborated upon in Section 10.11.

Secondly, it is of interest to monitor any significant changes that may occur from time to time in the nature of the spontaneous activity of a given neuron. One series of experiments has attempted to find a relationship between an animal's state of arousal and spontaneous neuronal activity (Evarts 1964; Lamarre et al. 1971; Benoit and Chataignier 1973; Steriade et al. 1973; Bassant 1976; Burns and Webb

1976; Webb 1976a, b). This has led to some unexpected findings. For example, the modal interspike interval is shorter for the sleep state in certain cells of the cat's visual and auditory cortices than it is in the waking or alarmed states (Webb 1976b). In other experiments the stability of spontaneous activity has been investigated during behavioral modification. O'Brien et al. (1973) found significant changes in the spontaneous activity of neurons in the postcruciate cortex of cats during classical conditioning, suggesting that detectable changes in the cells' electrophysiological properties had occurred. Ryan and Miller (1977) recorded the spontaneous activity of cells in the inferior colliculus of Rhesus monkeys. The mean discharge rates did not differ significantly during and "not-during" the performance of a reaction-time task.

Third, there may be systematic variations within structures. Correspondences between a neuron's location and features of its spontaneous activity were found in the cat cochlear nucleus (Pfeiffer and Kiang 1965) and in the monkey somatosensory cortex (Whitsel et al. 1972; Whitsel et al. 1977).

Analysis of a neuron's spontaneous activity is clearly an important prelude to the analysis of its *driven activity* in stimulus–response studies. The possibility has been raised that the background activity has significance as a carrier that is modulated by various stimuli (Rodieck et al. 1962). The ubiquitous noise within the nervous system, with particular reference to those parts that are involved in processing sensory data, has been noted by several authors (Levick et al. 1961; Siebert 1965).

In some first-order afferents the spike frequency is proportional to stimulus intensity, suggesting that cells are using a frequency code. But in most higher-order sensory cells, the discharge is irregular and the code is not at all obvious. This brings us to the important question (Moore, Perkel, and Segundo 1966) as to how the nervous system detects and identifies stimuli, a feat, which we must agree, it is remarkably good at. The nervous system must perform its own version of statistical inference in order to accomplish such identification. Does it make null hypotheses and reject or accept based on the evidence presented? There have been some intriguing schemes proposed for the processing of sensory inputs. The scheme proposed by Siebert (1965) for auditory signals and that for the decoding of spike trains in the somatosensory cortex in response to tactile stimulation (Schreiner et al. 1978) are particularly interesting. See also Holden (1982).

So we see there are two basic kinds of spike trains corresponding to (i) spontaneous and (ii) driven activity. We will mostly be concerned with analysis of the first type of activity simply because it has been better developed. An excellent earlier review was that of Perkel, Gerstein, and Moore (1967a).

Since the most common assumption made concerning the spike trains of spontaneously active cells is that they are *renewal processes*, we first define them and discuss their important properties insofaras they are relevant to neuronal activity. For more extensive treatments, see Cox and Miller (1965) and Karlin and Taylor (1975).

10.2   Renewal-process model of a spike train

Suppose an experiment begins at $t = 0$ and a sequence of spikes occurs at the random times $\Theta_1, \Theta_2, \Theta_3, \ldots$. The time intervals between successive spikes are $T_k = \Theta_k - \Theta_{k-1}$, $k = 1, 2, \ldots$, where $\Theta_0 \doteq 0$. We may count the number of spikes that occur up to and including time $t > 0$ and denote this by $N(t)$. It is assumed that spikes are distinguishable in time.

Definition
If the random variables T_k are independent and identically distributed, then $\{ N(t), t \geq 0\}$ is called a renewal process.

The name renewal process has industrial connotations. A machine or component is installed at $t = 0$. It breaks down or fails at time Θ_1, whereupon a replacement is installed. It is assumed that the time to failure of the second component has the same probability distribution as the original. The second component may eventually fail and is then replaced by a third, and so on. It is also supposed that the failure times of the various components are independent.

We see therefore that we may borrow techniques from *reliability theory* [see, for example, Barlow and Proschan (1975)] to analyze neuronal data. Furthermore, we may also draw on the methods employed in *survival analysis*, where failure times become the lifetimes of patients [see, for example, Lawless (1982) and Cox and Oakes (1983)].

Properties of renewal processes
The reader will have no trouble in convincing himself that a renewal process can be described in terms of either the sequence of variables $\{ T_k \}$, the sequence $\{ \Theta_k \}$, or the continuous family $\{ N(t) \}$.

In fact, we have

$$N(t) < k \quad \text{if an only if } \Theta_k > t, \qquad k = 1, 2, \ldots, \tag{10.1}$$

and

$$\Theta_k = T_1 + T_2 + \cdots + T_k, \qquad k = 1, 2, \ldots. \tag{10.2}$$

The relation (10.1) enables us to go back and forth between statements about the discrete variables $N(t)$ and the continuous variables Θ_k, since we must have

$$\Pr\{N(t) = k\} = \Pr\{\Theta_{k+1} < t\} - \Pr\{\Theta_k < t\}. \tag{10.3}$$

In this context, the T_k's are *interspike time intervals* (ISIs), Θ_k is the time of occurrence of the kth spike, and $N(t)$ is the number of action potentials up to and including time t. It follows, because of the assumed independence and identical distributions of the T_k's that, in the renewal model, the spike train is probabilistically specified by the common distribution function of the ISI

$$F(t) = \Pr\{T_1 \le t\}, \qquad t > 0, \tag{10.4}$$

or its density $f(t)$.

Other quantities that may be useful but all of which can be obtained from f or F are the following.

(i) *Spike-rate function.* The density of the interspike interval conditioned on nonappearance of a spike gives *the spike-rate function*

$$s(t) = \lim_{\Delta \to 0} \frac{\Pr\{t < T_1 \le t + \Delta t \mid T_1 > t\}}{\Delta t}. \tag{10.5}$$

In reliability studies this is called the failure-rate function, or hazard function, and was called the postimpulse probability by Poggio and Viernstein (1964).

From the definition of conditional probability (Section 9.2)

$$\Pr\{t < T_1 \le t + \Delta \mid T_1 > t\} = \frac{\Pr\{t < T_1 \le t + \Delta t\}}{\Pr\{T_1 > t\}}.$$

Hence we arrive at the relation

$$s(t) = f(t) / [1 - F(t)]. \tag{10.6}$$

The quantity $s(t)$ indicates how imminent a spike is.

(ii) *Spike density.* In the theory of renewal processes, the *renewal density* is defined as

$$u(t) = \lim_{\Delta t \to 0} \frac{\Pr\{\text{an event occurs in } (t, t + \Delta t]\}}{\Delta t}. \tag{10.7}$$

For spike trains we call this the *spike density*. It will be seen that $u(t)$ is also the rate of change of the expected number of spikes in $(0, t]$,

$$u(t) = d/dt\left[E(N(t))\right]. \tag{10.8}$$

We now prove that $u(t)$ is given by

$$u(t) = \sum_{k=1}^{\infty} f_k(t), \tag{10.9}$$

where f_k is the probability density function of Θ_k, the waiting time for the kth spike.

Proof. Let the distribution function of Θ_k be $F_k(t)$. Then, from the definition of expectation and (10.3),

$$E(N(t)) = \sum_{k=1}^{\infty} k \Pr\{N(t) = k\}$$

$$= \sum_{k=1}^{\infty} k\left[F_k(t) - F_{k+1}(t)\right]$$

$$= \sum_{k=1}^{\infty} kf_k(t) - \sum_{k=1}^{\infty} (k+1)F_{k+1}(t) + \sum_{k=1}^{\infty} F_{k+1}(t)$$

$$= \sum_{k=1}^{\infty} F_k(t).$$

Differentiating, we obtain (10.9).

Now from (10.9) we see that

$$u(t)\,\Delta t = \sum_{k=1}^{\infty} \Pr\{k\text{th spike in } (t, t+\Delta t]\}$$

$$= \Pr\{\text{a spike occurs in } (t, t+\Delta t]\}, \tag{10.10}$$

which agrees with (10.7).

It is seen from (10.9) that *the spike density $u(t)$ is the sum of the probability density functions for the waiting times for the first, second,... spikes.* We will see therefore that $u(t)$ can be derived from the common density $f(t)$ of the ISIs.

Let X and Y be any two continuous random variables taking positive values. Let their probability densities be f_X and f_Y, respectively. Suppose further that X and Y are independent and that we obtain a third random variable Z as their sum:

$$Z = X + Y. \tag{10.11}$$

Then [see, for example, Tuckwell (1988a)] the probability density of Z

is given by the convolution integral

$$f_Z(z) = \int_0^\infty f_X(x) f_Y(z-x)\, dx. \tag{10.12}$$

Since $\Theta_2 = T_1 + T_2$ and T_1 and T_2 have common density f,

$$f_2(t) = \int_0^\infty f(v) f(t-v)\, dv.$$

Similarly, $\Theta_3 = \Theta_2 + T_3$ so the probability density of Θ_3 is the convolution of f_2 and f_1 and so forth. In principle, therefore, $u(t)$ can always be found from $f(t)$ if the renewal model is valid.

Examples

(a) *Poisson Process.* If the waiting times between spikes are exponential random variables with common density

$$f(t) = \lambda e^{-\lambda t}, \qquad t > 0,$$

then $\{N(t), t \geq 0\}$ is a simple Poisson process with intensity λ. It is left as an exercise to show that

$$s(t) = u(t) = \lambda,$$

so that *both the spike-rate function and the spike density are constant.*

(b) *The ISI is gamma-distributed.* Let the ISIs have the common density

$$f(t) = \frac{\lambda(\lambda t)^{m-1} e^{-\lambda t}}{(m-1)!}, \qquad t > 0,$$

where m is a strictly positive integer and $\lambda > 0$. Then since each T_k can be viewed as the waiting time for m events in a Poisson process, we find

$$s(t) = \frac{\lambda(\lambda t)^{m-1}}{(m-1)!\left(1 + \lambda t + \dfrac{(\lambda t)^2}{2!} + \cdots + \dfrac{(\lambda t)^{m-1}}{(m-1)!}\right)}, \tag{10.13}$$

$$u(t) = \lambda e^{-\lambda t} \sum_{k=1}^\infty \frac{(\lambda t)^{km-1}}{(km-1)!}. \tag{10.14}$$

In this case the function $u(t)$ is the sum of gamma densities with parameters λ and $m, 2m, 3m, \ldots$. Hence $u(t)$ will have a first peak roughly at the maximum of f_1, then another near the maximum of f_2, and so on. This gives $u(t)$ the appearance of a damped sinusoid as is often observed experimentally, as, for example, in ventrobasal thalamic neurons (Poggio and Viernstein 1964). Syka et al. (1977) have given spike densities for cells of the rat mesencephalic reticular formation but $u(t)$ is called the *autocorrelation* as it was in the pioneering work of Gerstein and Kiang (1960). Some pictures are given in Figure 10.5 of histograms that estimate $u(t)$ (see p. 243).

(iii) *Asymptotic properties as $t \to \infty$.* Let $\{N(t)\}$ be a renewal process as above and let $E(T_1) = \mu$ and $\text{Var}(T_1) = \sigma^2$. The variables $N(t)$ have the following properties as $t \to \infty$, as is proved, for example, in Cox (1962) and Cox and Miller (1965).

(a) *Asymptotic normality.* A result of applying the central limit theorem is that the number of spikes in $(0, t]$ is approximately normally distributed with mean t/μ and variance σ^2/μ^3. That is,

$$N(t) \overset{d}{\underset{t \to \infty}{\to}} N\left(\frac{t}{\mu}, \frac{\sigma\sqrt{t}}{\mu^{3/2}} \right), \tag{10.15}$$

where d signifies pointwise convergence of the distribution function.

(b) *Approach to the asymptotic mean and variance.* For large t we have, asymptotically,

$$E(N(t)) \underset{t \to \infty}{\sim} \frac{t}{\mu} + \frac{\sigma^2 - \mu^2}{2\mu^2} + o(1), \tag{10.16}$$

$$\text{Var}(N(t)) \underset{t \to \infty}{\sim} \frac{\sigma^2 t}{\mu^3} + \left(\frac{1}{12} + \frac{5\sigma^4}{4\mu^4} - \frac{2\mu_3}{3\mu^3} \right) + o(1), \tag{10.17}$$

where $\sim$ signifies that the ratio of the quantities on the left-hand side to those on the right-hand side approaches 1 as $t \to \infty$, here $o(1)$ represents terms that go rapidly (like e^{-t}) to zero and μ_3 is the third central moment, $\mu_3 = E((T_1 - \mu)^3)$. A simple consequence of (10.16) is that

$$u(t) \underset{t \to \infty}{\sim} 1/\mu. \tag{10.18}$$

From (10.16) and (10.17), we have

$$\frac{\text{Var}(N(t))}{E(N(t))} \sim \frac{\sigma^2}{\mu^2}. \tag{10.19}$$

The relation (10.18) indicates that the spike density becomes constant for large t, no matter what the density of the ISI.

The model we have described, in which a spike occurs at $t = 0$, is called an *ordinary renewal process* and in this case $T_1, T_2, \ldots$ all have the same distribution. It may happen that the observation of a spike train does not commence with a spike so that the waiting time to the first spike T_1 has a different distribution from subsequent intervals $T_2, T_3, \ldots$. The process so obtained is called a *modified renewal process* and we suppose for it that T_1 has density $f_1(t)$, whereas $T_2, T_3, \ldots$ have common density $f(t)$.

There is a third model, called an *equilibrium renewal process*, which occurs as a special case of the modified renewal process. The density of T_1 is given in this special case by

$$f_1(t) = [1 - F(t)]/\mu.$$

The reason for this special choice is that it is the asymptotic $(t \to \infty)$ density of the waiting time to a spike from an arbitrary time in an ordinary or modified renewal process. Thus if we start with a waiting time for a spike given by $f_1(t)$, it is as if the process has been running for an infinite time. Considering that neurons when first observed in the laboratory have usually been spiking for a very long time, it might seem that the equilibrium renewal process is appropriate in many circumstances. However, we can always choose the time origin to start at a spike and then employ the ordinary renewal-process model provided the interspike intervals are i.i.d. random variables for large t, no matter what the density of the ISI.

10.3 Estimation of the parameters of an ISI distribution

Let $\{T_1, T_2, \ldots, T_n\}$ be a sequence of interspike time intervals. If these satisfy the renewal hypothesis, they form a set of independent and identically distributed random variables, which (without regard to order) constitutes a random sample for an underlying population variable T whose mean and variance are μ and σ^2 (the same quantities as in the previous section). The mean ISI is the *sample mean*

$$\bar{T} = \frac{T_1 + T_2 + \cdots + T_n}{n}, \quad \text{with value } \bar{t} = \frac{t_1 + t_2 + \cdots + t_n}{n}.$$

(10.20)

The sample variance is

$$S_T^2 = \frac{1}{n-1} \sum_{k=1}^{n} (T_k - \bar{T})^2, \quad \text{with value } s_T^2 = \frac{1}{n-1} \sum_{k=1}^{n} (t_k - \bar{t})^2.$$

(10.21)

As shown in basic statistics texts [e.g., Walpole and Myers (1985)],

$$E(\bar{T}) = \mu,$$ (10.22)

$$E(S_T^2) = \sigma^2.$$ (10.23)

Thus $\bar{T}$ and S_T^2 provide *unbiased* estimates for μ and σ^2. The mean ISI is, in fact, approximately normally distributed for large n, with standard deviation $\sigma/\sqrt{n}$, regardless of the nature of the distribution of the underlying population (central limit theorem). Note that in spike-train experiments, n is usually on the order of thousands, though there are problems with stationarity.

If one assumes that the ISI distribution has a given form (e.g., normal with mean μ and variance σ^2), then the parameters of the distribution can be estimated from the data by various methods. Two such methods, which often yield analytical results, are the *method of moments* and the *maximum-likelihood* estimation. Both of these will be illustrated in the discussion that follows. We will examine a few cases in detail, but these problems are very well dealt with elsewhere in other contexts [see, for example, Lawless (1982)]. We present a classification scheme in Section 10.12.

(a) *Exponential distribution*

Exponentially distributed intervals are important in connection with Poisson processes, tests for which are described below. They have been fitted to neuronal ISI histograms of cells of the cochlear nucleus (Rodieck et al. 1962), the ventrolateral nucleus of the thalamus (Lamarre et al. 1971), pyramidal tract cells (Steriade et al. 1973), and primary afferents in the vestibular system (Correia and Landolt 1977).

(i) *Method of moments.* The parameters (in this case only one) can sometimes be estimated by equating the moments of the hypothesized distribution to those estimated from the sample. Since $E(T) = 1/\lambda$, this gives, using a caret for an estimator with obvious notation,

$$\hat{\Lambda} = 1/\bar{T}, \quad \text{with value } \hat{\lambda} = 1/\bar{t}. \tag{10.24}$$

(ii) *Maximum-likelihood estimates.* A likelihood function measures the chance of a given set of observations under a given hypothesis. The likelihood function in the present context is defined as

$$L(\lambda) = \prod_{k=1}^{n} f(t_k; \lambda)$$

$$= \lambda^n \exp\left(-\lambda \sum_{k=1}^{n} t_k\right). \tag{10.25}$$

The estimated value of λ is that which maximizes the likelihood function. [Note that more accurately the t_k's in (10.25) should be replaced by the T_k's, which are random variables, as is therefore L. The present notation is commonly employed.] It is often easier to utilize the fact that the logarithmic function is monotonic so that we may find $\hat{\lambda}$ by maximizing $\ln L(\lambda)$. We have

$$\frac{d}{d\lambda}\ln L(\lambda) = \frac{d}{d\lambda}\left(n \ln \lambda - \lambda \sum_{k=1}^{n} t_k\right)$$

$$= \frac{n}{\lambda} - \sum_{k=1}^{n} t_k.$$

Setting this equal to zero to obtain a maximum (the second derivative is $-n/\lambda^2$, which must be negative), we obtain

$$\hat{\Lambda} = 1/\overline{T}, \quad \text{with value } \hat{\lambda} = 1/\overline{t}.$$

This is the same as the estimate from the method of moments.

Inclusion of an absolutely refractory period

If there is an absolute refractory period following each spike and its duration is t_R, then the ISI density is modified to

$$f(t; \lambda, t_R) = \begin{cases} 0, & 0 < t < t_R, \\ \lambda e^{-\lambda(t-t_R)}, & t > t_R. \end{cases} \tag{10.26}$$

The mean and variance are $1/\lambda + t_R$ and $1/\lambda^2$, respectively. The method of moments therefore yields the relations

$$\hat{\Lambda} = 1/S_T,$$

$$\hat{T}_R = \overline{T} - S_T.$$

A random variable with the density (10.26) is said to have a two-parameter exponential distribution. It has been employed as an ISI density by Poggio and Viernstein (1964) and Correia and Landolt (1977).

Binned data

Suppose the various t_k's have been binned with a constant bin width of Δt. The ith bin will include ISIs satisfying

$$(i-1)\Delta t < t \leq i \Delta t, \quad i = 1, 2, \ldots, m. \tag{10.27}$$

The probability that any observation from the exponential density falls in the ith bin is

$$\lambda \int_{(i-1)\Delta t}^{i\Delta t} e^{-\lambda t}\, dt = e^{-\lambda(i-1)\Delta t}\left(1 - e^{-\lambda \Delta t}\right). \tag{10.28}$$

If there are n_i observations in the ith bin, the likelihood function is

$$L(\lambda) = \prod_{i=1}^{m} e^{-\lambda n_i(i-1)\Delta t}\left(1 - e^{-\lambda \Delta t}\right)^{n_i}.$$

On taking logarithms,

$$\ln L(\lambda) = -\lambda \sum_{i=1}^{m} n_i(i-1)\, \Delta t + n_T(1 - e^{-\lambda \Delta t}), \qquad (10.29)$$

where

$$n_T = \sum_{i=1}^{m} n_i \qquad (10.30)$$

is the total number of spikes in the train.

It is left as an exercise to show that the value of the maximum-likelihood estimate of λ is

$$\hat{\lambda} = \frac{1}{\Delta t} \ln\left(1 + \frac{n_T}{\sum n_i(i-1)}\right). \qquad (10.31)$$

This estimate was employed by Smith and Smith (1965) for spontaneously active cells in the cat cerebral cortex.

(b) Gamma densities

Let the ISI density be

$$f(t; \lambda, m) = \frac{\lambda(\lambda t)^{m-1} e^{-\lambda t}}{(m-1)!}, \qquad t > 0,$$

where we are assuming m is an integer, though this is not necessary. Using the method of moments, one easily gets

$$\hat{m}/\hat{\lambda} = \bar{t}, \qquad (10.32A)$$

$$\hat{m}/\hat{\lambda}^2 = s_T^2. \qquad (10.32B)$$

Hence we may estimate λ and m with the random variables

$$\hat{\Lambda} = \bar{T}/S_T^2, \qquad (10.33A)$$

$$\hat{M} = \bar{T}^2/S_T^2. \qquad (10.33B)$$

The maximum-likelihood estimates for λ and m are not obtained so easily. However, it may be shown that

$$\frac{\hat{m}}{\hat{\lambda}} = \bar{t}, \qquad (10.34A)$$

$$\ln \hat{m} - \psi(\hat{m}) = \ln(\bar{t}/\tilde{t}), \qquad (10.34B)$$

where $\tilde{t}$ is the geometric mean interval

$$\tilde{t} = \left(\prod_{k=1}^{n} t_k\right)^{1/n}, \qquad (10.35)$$

and $\psi(x) = \Gamma'(x)/\Gamma(x)$ is the digamma function. Equation (10.34B) must be solved numerically. Some approximations are given in Lawless (1982).

ISI histograms are often found with the shape of a gamma density. Examples where gamma densities have been fitted to ISI data are retinal ganglion cells (Bishop et al. 1964), red nucleus cells (Nakahama et al. 1968), cells of the somatosensory cortex, and cells of the vestibular system (Correia and Landolt 1977). The latter authors used least-squares curve fitting to obtain parameter values.

(c) *Normal random variables*

In the context of ISI densities, gamma densities with large values of m are probably more appropriate than the normal distributions that they approximate. However, some ISI densities resemble normal densities very closely. Examples are found in the spontaneous activity of some cells of the cat cochlear nucleus (Rodieck et al 1962; Pfeiffer and Kiang 1965). The normal density, when used for an ISI, has the major defect of allowing negative values. If, however, the probability mass is concentrated at positive values, this will be of little consequence.

For the normal density with mean μ and variance σ^2,

$$f(t;\mu,\sigma^2) = \frac{1}{\sqrt{2\pi\sigma^2}}\exp\left(-\frac{(t-\mu)^2}{2\sigma^2}\right), \qquad -\infty < t < \infty,$$

(10.36)

the method of moments and maximum-likelihood estimates are

$$\hat{\mu} = \bar{t},$$
$$\hat{\sigma}^2 = \frac{1}{n}\sum_{k=1}^{n}(t_k - \bar{t})^2.$$

(10.37)

Alternatively, s_T^2 can be used to estimate σ^2. These results are useful in the following case.

(d) *Lognormal distribution*

The lognormal distribution has been used by Burns and Webb (1976) to characterize the random discharges of cells in the cat cerebral cortex. The claim that only two parameters are needed to describe an entire spike train is attractive, but, of course, assumes ISIs are independent. If T is lognormal, then by definition $\ln T$ is normal. The density of T is

$$f(t;\mu,\sigma^2) = \frac{1}{t\sqrt{2\pi\sigma^2}}\exp\left\{-\frac{(\ln t - \mu)^2}{2\sigma^2}\right\}, \qquad t > 0.$$

(10.38)

To find the parameters μ and σ^2 it is advisable to convert the data to a histogram for $\ln T$ – then just employ the estimates (10.37) for normal random variables to find μ and σ^2.

(e) First-passage time for a Wiener process

The density of the first-passage time of a Wiener process, with drift $\mu > 0$ and variance parameter σ^2, to a barrier at θ when the initial position is $x_0 < \theta$ is given in (9.99). This distribution is often called *inverse Gaussian* [see Folks and Chhikara (1978) for a review]. The gamma density is obtained as the ISI when the membrane potential is purely discontinuous (i.e., Poisson process) in the case of a cell with an infinite time constant. Similarly, the density of (9.99) is obtained when the potential is purely continuous and there is no decay. One would not expect either of these zero-decay models to be appropriate but there has been a persistent effort to use them. The reason for the thriving of this inadequate model is clearly that there are no closed-form expressions for the densities in the more realistic Stein model or the Ornstein–Uhlenbeck process, which are obtained when the exponential decay of membrane potential between inputs is included.

An attempt to find the parameters by the method of moments using

$$E(T) = \frac{\theta - x_0}{\mu},$$

$$\text{Var}(T) = \frac{(\theta - x_0)\sigma^2}{\mu^3}, \tag{10.39}$$

$$E\left((T - E(T))^3\right) = \frac{3(\theta - x_0)\sigma^4}{\mu^5}, \quad \text{etc.}$$

will fail because these are functions of the variables $(\theta - x_0)/\mu$ and σ/μ. That is, only two independent parameters are required to specify the density. This can be traced to the fact that the time of first passage of

$$X(t) = x_0 + \mu t + \sigma W(t)$$

to θ, is the same as the time of first passage of

$$Y(t) = \frac{X(t) - x_0}{\mu} = t + \frac{\sigma}{\mu} W(t) \tag{10.40}$$

to $((\theta - x_0)/\mu)$ from $Y(0) = 0$. Introducing

$$\alpha = \frac{\theta - x_0}{\mu},$$

$$\beta = \frac{\sigma}{\mu}, \tag{10.41}$$

the moments become

$$E(T) = \alpha,$$
$$\text{Var}(T) = \alpha\beta^2. \tag{10.42}$$

This leads to straightforward estimates by the method of moments

$$\hat{\alpha} = \bar{t},$$
$$\hat{\beta} = s_T^2/\bar{t}. \tag{10.43}$$

Maximum-likelihood estimates

In view of the fact that there are only two independent parameters α and β as given by (10.41), we find estimators for these by maximizing a likelihood function. Put

$$f(t; \alpha, \beta) = \frac{\alpha}{\sqrt{2\pi\beta^2 t^3}} \exp\left(-\frac{(\alpha - t)^2}{2\beta^2 t}\right). \tag{10.44}$$

The likelihood function is

$$L(\alpha, \beta) = \left(\frac{\alpha}{\beta}\right)^n (2\pi)^{-n/2} \prod_{k=1}^{n} \frac{\exp\left(-\dfrac{(\alpha - t_k)^2}{2\beta^2 t_k}\right)}{t_k^{3/2}}. \tag{10.45}$$

Taking logarithms and differentiating gives

$$\frac{\partial \ln L}{\partial \alpha} = \frac{n}{\alpha} - \sum_{k=1}^{n} \frac{\alpha - t_k}{\beta^2 t_k},$$
$$\frac{\partial \ln L}{\partial \beta} = -\frac{n}{\beta} + \sum_{k=1}^{n} \frac{(\alpha - t_k)^2}{\beta^3 t_k}. \tag{10.46}$$

Setting the partial derivatives equal to zero and solving for α and β,

$$\hat{\alpha} = \bar{t},$$
$$\bar{\beta} = \sqrt{\bar{t}\left(-1 + \frac{\bar{t}}{n} \sum_{k=1}^{n} \frac{1}{t_k}\right)}. \tag{10.47}$$

Lansky (1983) has a slightly different parameterization of f. The density (10.44) has been fitted to neuronal interspike time data by Gerstein and Mandelbrot (1964), and see Figure 9.15, by Nilsson (1977), and by Correia and Landolt (1977). Nilsson compared results using maximum-likelihood estimators and least squares and concluded that the latter was the better method. Thus moment or maximum-likelihood estimates are useful to provide initial guesses for least-squares routines.

However, estimates of the parameters of the density (10.44) are of little physiological interest because the underlying model cannot be taken seriously. They may, however, be useful in comparative studies.

(f) *Densities for more realistic models*

Because of the paucity of analytical results for the more realistic models, which include the decay of potential between inputs, reversal potentials, and spatial extent, there has been no systematic investigation of parameter investigation. Clearly, new techniques are needed. Tuckwell and Richter (1978) used the method of moments to obtain estimates of the parameters in Stein's model for cells of the cat cochlear nucleus. The resulting estimates were very reasonable.

10.4   The nature of statistical tests

There are numerous excellent introductory texts on statistics [for example, Walpole and Myers (1985)]. The following few paragraphs are just to remind the reader concerning the nature of statistical tests and to name a few of the more common ones.

In testing the validity of a stochastic ($\equiv$ random, probabilistic) model, it is often necessary to perform statistical tests on data. For example, in the renewal-process model outlined in Section 10.2, one hypothesis that is made is that successive ISIs are independent. We would call a particular hypothesis such as this a *null hypothesis H_0*. One may construct random variables with known, or approximately known, distributions under the assumption of independence. One examines the observed values of such random variables, which are called *test statistics*, or just *statistics*. If the probability of occurrence of such values is judged to be very small under the independence hypothesis, one is inclined to reject that hypothesis.

The probability of values of a test statistic as extreme or more so than observed, under H_0, is called the *significance probability*, or the observed significance level or *P*-value. Values of P less than 0.05 are usually taken as strong evidence that H_0 is unlikely to be true.

We have already defined the terms *random sample*, *sample mean*, and *sample variance* in the preceding section. We will now briefly consider the most common random variables that arise as test statistics in the present context. These are the standard normal, the *t*-statistic, and χ^2.

(a) *Standard normal*

The usual symbol is Z or $N(0, 1)$ to indicate that the mean is zero and the standard deviation 1 [sometimes the second argument in

$N(\cdot,\cdot)$ is the variance]. The probability density is

$$f_Z(x) = \frac{1}{\sqrt{2\pi}} \exp\left(-\frac{x^2}{2}\right), \qquad -\infty < x < \infty. \qquad (10.48)$$

Tables of $z_{\alpha/2}$, defined through

$$\Pr\{Z > z_{\alpha/2}\} = \alpha/2, \qquad (10.49)$$

are available in most statistics texts.

(b) *t-statistic*

If $\{X_1, X_2, \ldots, X_n\}$ is a random sample of size n from a normal distribution with mean μ and standard deviation σ, the sample mean is

$$\overline{X} = \frac{1}{n} \sum_{k=1}^{n} X_k, \qquad (10.50)$$

and the sample variance is

$$S^2 = \frac{1}{n-1} \sum_{k=1}^{n} (X_k - \overline{X})^2. \qquad (10.51)$$

The random variable

$$\mathcal{T}_n = \frac{\overline{X} - \mu}{S/\sqrt{n}} \qquad (10.52)$$

is said to have a *t*-distribution with parameter $n-1$ (or $n-1$ degrees of freedom). (One would usually use T_n or T for this variable, but $\mathcal{T}_n$ avoids confusion with our symbol for an ISI.) The density of $\mathcal{T}_n$ is

$$f_{\mathcal{T}_n}(x) = \frac{\Gamma\left(\dfrac{n}{2}\right)}{\sqrt{\pi(n-1)}\;\Gamma\left(\dfrac{n-1}{2}\right)} \left(1 + \frac{x^2}{n-1}\right)^{n/2}, \qquad -\infty < x < \infty.$$

$$(10.53)$$

Like the density of Z, that of $\mathcal{T}_n$ is symmetric about $x = 0$, but has larger tails. Tables of values of $t_{\alpha/2,\,n-1}$ defined by

$$\Pr\{\mathcal{T}_n > t_{\alpha/2,\,n-1}\} = \alpha/2, \qquad (10.54)$$

can also be found in most statistics texts.

(c) χ^2-*variable*

Let $\{X_1, X_2, \ldots, X_n\}$ be a random sample from a standard normal distribution. That is, each X_k has the density (10.48) and the collection of variables is an independent set. Then the sum of their squares

$$\chi_n^2 = \sum_{i=1}^{n} X_i^2 \qquad\qquad (10.55)$$

is called a χ^2-random variable with n degrees of freedom. The density of χ_n^2 is

$$f_{\chi_n^2}(x) = \frac{1}{2^{n/2}\Gamma(n/2)} x^{n/2-1} e^{-x/2}, \qquad x > 0. \qquad (10.56)$$

The χ^2-variable arises frequently in goodness-of-fit tests as follows. A random experiment is performed and observations may fall into any of n distinct categories. Under a null hypothesis H_0, suppose that the probability is p_i that any observation falls in category i. If there are N observations altogether, the expected number in category i is Np_i and this may be compared with the number N_i (with observed values n_i), which actually fall in category i when the experiment is performed. To compare the hypothesized and observed situations, we compute the value of the random variable

$$D_n^2 = \sum_{i=1}^{n} \frac{(N_i - Np_i)^2}{Np_i} \overset{d}{\simeq} \chi_{n-1}^2. \qquad (10.57)$$

When N is large, D_n^2 has approximately the same distribution as χ_{n-1}^2. Large observed values of the χ^2-statistic make us suspect that H_0 is false. Critical values $\chi_{\alpha;\,n-1}^2$ are defined through

$$\Pr\left\{\chi_{n-1}^2 > \chi_{\alpha,\,n-1}^2\right\} = \alpha, \qquad (10.58)$$

and are given in most statistics texts.

10.5 Testing and comparing distributions

The following are some situations in which testing distributions is important.

(i) Goodness-of-fit test to a given distribution

Suppose a particular distribution (e.g., gamma, normal, etc.) has been fitted to an ISI histogram with parameters estimated by one of the methods outlined in Section 10.3. It is then required to see how well the data fit the calculated distribution. The tests that are then employed are usually nonparametric goodness-of-fit tests of which the following two are the most common.

(a) χ^2-goodness-of-fit test. The theory of the χ^2-test was outlined in the previous section. The general rule is that the number of degrees of freedom of χ^2 is always reduced by one since the relation $\sum n_i = N$ must hold. A further degree of freedom is lost for each parameter estimated from the data. Using χ^2-tests, ISI histograms have been tested against exponential (Smith and Smith 1965), gamma (Nakahama et al. 1968), and lognormal distributions (Burns and Webb 1976). The latter authors, incidentally, obtained the parameters of the distribution by minimizing χ^2.

(b) *Kolmogorov–Smirnov test.* Suppose the interspike time interval is hypothesized to have distribution function $F_0(t)$, $t > 0$. To test whether a set of observations on $\{T_k, k = 1, 2, \ldots, n\}$ of the ISI is likely to come from such a distribution, we construct an empirical distribution function

$$\tilde{F}_n(t) = \frac{\text{number of the } T_k\text{'s less than or equal to } t}{n} \qquad (10.59)$$

This may be represented as

$$\tilde{F}_n(t) = \frac{1}{n} \sum_k H(t - T_{(k)}), \qquad (10.60)$$

where $H(\cdot)$ is a unit step function and $T_{(k)}$ are the order statistics $T_{(1)} < T_{(2)} < \cdots < T_{(n)}$. This makes it clear that F_n has a jump of magnitude $1/n$ at each $T_{(k)}$. To examine how closely the empirical and hypothetical distribution functions agree, we compute the greatest value of the absolute value of their difference. This gives the Kolmogorov–Smirnov statistic

$$D_n = \sup_{t>0} \left| \tilde{F}_n(t) - F_0(t) \right|. \qquad (10.61)$$

Large values of D_n render F_0 an unlikely candidate as a distribution function. Note that the random variable D_n does not depend on F_0 – the test is distribution-free. Tables of critical values of D_n for various n are available in many texts [see, for example, Pearson and Hartley (1972)].

The Kolmogorov–Smirnov test has been used as a goodness-of-fit test for ISI data by Whitsel et al. (1972) and Correia and Landolt (1977). Note that there are superior tests available if one is testing against certain particular kinds of distribution functions, such as exponential [see Lawless (1982), Chapter 9].

(ii) *Detecting changes in the ISI*

It is often of interest to compare the firing pattern of a neuron at different times or in different physiological, pharmacological, or behavioral states. If an empirical distribution function $F_{n_1}(t)$ is obtained in one set of circumstances with n_1 observations of the ISI, and $F_{n_2}(t)$ is obtained with n_2 observations in another set of circumstances, a comparison can be made by computing the value of the two-sample Kolmogorov–Smirnov statistic. That is,

$$D_{n_1, n_2} = \sup_{t > 0} \left| \tilde{F}_{n_1}(t) - \tilde{F}_{n_2}(t) \right|. \tag{10.62}$$

Critical values for D_{n_1, n_2} are given in Pearson and Hartley (1972).

The Kolmogorov–Smirnov test was employed by Pyatigorski (1966) to compare ISIs of Clarke's column cells before and after the administration of anaesthetic. Each sample was obtained from three minutes of spike records. It was also employed by Stein and Matthews (1965) to compare the firing rates of muscle spindles during different time intervals in order to test for stationarity. Of course, sometimes changes are so obvious that no formal tests are necessary [see, for example, Syka et al. (1977)].

A test that may be used for comparing several samples is the nonparametric Kruskal–Wallis test. The whole set of $n = n_1 + n_2 + \cdots + n_m$ ISIs is arranged in ascending order and a rank assigned $(1, 2, \ldots, n)$ to each observation. Let $R_1, R_2, \ldots, R_m$ be random variables that are the sums of the $n_1, n_2, \ldots, n_m$ ranks, respectively, for the m samples. Then the Kruskal–Wallis H-statistic is

$$H = \frac{12}{n(n+1)} \sum_{k=1}^{m} \frac{R_k^2}{n_k} - 3(n+1). \tag{10.63}$$

If the null hypothesis that the samples come from the same distribution is true, then H has approximately a χ^2-distribution with $m - 1$ degrees of freedom. In the case of $m = 2$ the test is called a *Wilcoxon test*.

Less sensitive tests can be made on such quantities as the mean ISI in different time periods (Velikaya and Kulikov 1966), or under different conditions such as sleep versus waking (Evarts 1964; Bassant 1976; Burns and Webb 1976), or different behavioral states (Ryan and Miller 1977).

If the ISI has approximately a normal distribution, one may test the equality of the means of two samples of sizes n_1 and n_2, with sample means $\overline{T}_1$ and $\overline{T}_2$ and sample variances $S_{T_1}^2$ and $S_{T_2}^2$, by computing the value of the statistic

$$\mathcal{T}_\nu = \frac{\overline{T}_1 - \overline{T}_2}{\sqrt{\left(S_{T_1}^2/n_1\right) + \left(S_{T_2}^2/n_2\right)}}. \tag{10.64}$$

Under the null hypothesis that the means are equal, this statistic has approximately a t-distribution with

$$\nu = \frac{\left[S_{T_1}^2/n_1 + S_{T_2}^2/n_2 \right]^2}{\dfrac{\left(S_{T_1}^2/n_1 \right)^2}{n_1 - 1} + \dfrac{\left(S_{T_2}^2/n_2 \right)^2}{n_2 - 1}} \qquad (10.65)$$

degrees of freedom (Walpole and Myers 1985). Thus inequality of the means (and hence the distributions) will be supported by large observed values of $\mathcal{T}_\nu$. For large enough n_1 and n_2, $\mathcal{T}_\nu$ can be replaced by Z.

10.6 Stationarity

One of the important steps in the stochastic analysis of a spike train or set of similar neural data, is to ascertain whether the underlying process is stationary, as in the case of an equilibrium renewal process.

A *counting process* such as $\{ N(t), t \geq 0 \}$, where $N(t)$ is the number of spikes in $(0, t]$, is called *stationary in the narrow sense* if the joint probability distribution of the numbers of spikes in any collection of subsets of the space of time points is invariant under time translations. Alternative expressions are *strictly or strongly stationary*. A process is called *wide-sense stationary* if its mean and covariance functions are invariant under time translations. Alternative terms are *covariance, second-order, Khinchin, or weakly stationary*.

A quick, but crude, method for testing for stationarity is to examine records of the mean discharge rate versus time. A steady mean rate is a necessary condition for stationarity. Plots called *tachograms* in which the length of each interval is plotted against its serial number, as in Figure 9.1, may also be useful (Hagiwara 1954; Junge and Moore 1966). Note that *adaptation* is noticeable as a monotonic increase in the mean interval and this is an often encountered form of nonstationarity in spike trains.

A better approach is to perform statistical tests as outlined in the previous section on the mean ISI from different parts of the record. An even more rigorous approach consists of comparing the empirical distributions of the ISI over different time intervals, using, for example, the Kolmogorov–Smirnov statistic as outlined previously. Tests for stationarity in the narrow sense are not generally employed.

One test that has been employed for testing wide-sense stationarity of neural data is the nonparametric *runs test*. To execute this, the total time period of observation is partitioned into, say n, equal subintervals with mean intervals $\bar{T}_1, \bar{T}_2, \ldots, \bar{T}_n$. The *median* of the averages is found and then the average for each subinterval is scored as $+$ if it is above the median and $-$ if below. The total number of $+$'s, n_1 say, and of $-$'s, n_2 say, are recorded with total $n = n_1 + n_2$. A *run* is an unbroken sequence of one or more $+$'s or $-$'s. If the probabilistic structure of the process is not changing, $+$ and $-$ are equally likely.

The total number of runs is a random variable V. Large or small observed values of V lead to the rejection of the null hypothesis H_0 of weak stationarity: large values because they represent the occurrence of alternating large and small mean intervals in a predictable fashion; small values lead to rejection of H_0 because it is unlikely that the mean will remain above the median of the means in several consecutive time intervals. The probability distribution of V for various n_1 and n_2 is tabulated [see, for example, Walpole and Myers (1985)] and hence we may test H_0. For large n_1 and n_2 (> 10), under H_0, the variable

$$Z = (V - \mu_V)/\sigma_V \tag{10.66}$$

has approximately a standard normal distribution, where

$$\mu_V = \frac{2n_1 n_2}{n_1 + n_2} + 1, \tag{10.67}$$

$$\sigma_V = \frac{2n_1 n_2 (2n_1 n_2 - n_1 - n_2)}{(n_1 + n_2)^2 (n_1 + n_2 - 1)}. \tag{10.68}$$

Further tests for stationarity in connection with renewal and Poisson processes are outlined in Sections 10.8 and 10.9.

Cumulative sums

A graphical method for detecting departures from stationarity has been advocated by Muschaweck and Loevner (1978). This is called a plot of the *cumulative sum* (abbreviated to *cusum*) defined as follows. Let $T_1, T_2, \ldots, T_n$ be a sequence of ISIs. Then the cusum at r is the random variable

$$Q_r = r\left[\frac{1}{r}\sum_{k=1}^{r} T_k - \frac{1}{n}\sum_{k=1}^{n} T_k\right], \tag{10.69}$$

where $r = 1, 2, \ldots$ is the serial number of a spike. Thus Q_r is r times the difference between the mean of the first r ISIs and the grand mean. If the process is stationary in the wide sense, we will have

$$E(Q_r) = 0, \qquad r = 1, 2, \ldots, n. \tag{10.70}$$

Suppose, however, that the population mean interval is μ_1 for the first m ISIs and μ_2 for the remaining $n - m$. The grand mean is

$$\mu = \frac{m}{n}\mu_1 + \left(\frac{n-m}{n}\right)\mu_2.$$

We also have

$$E(Q_r) = r(\mu_1 - \mu), \qquad r = 1, 2, \ldots, m, \tag{10.71}$$

whereas, after the change in distribution,

$$E(Q_r) = m\mu_1 + (r - m)\mu_2 - \mu r, \qquad r = m+1, \ldots, n. \tag{10.72}$$

Hence the slope of $E(Q_r)$ versus r changes from

$$dE(Q_r)/dr = \mu_1 - \mu \tag{10.73}$$

to

$$dE(Q_r)/dr = \mu_2 - \mu. \tag{10.74}$$

A change in distribution should therefore be easy to detect graphically.

Muschawek and Loevner used this technique to select portions of a spike train, which could be regarded as stationary and further analyzed. The cusum technique is the subject of a recent investigation of how to detect a change in the drift parameter of a Wiener process (Pollak and Siegmund 1985). A further reference is the book of van Dobben de Bruyn (1968).

10.7 Independence of ISIs

If a spike train is to be modeled as renewal process, then the interspike intervals must be *independent*. The nature of spike generation makes it very unlikely that this independence assumption would be satisfied, especially in rapidly firing cells. Each impulse has an aftereffect on subsequent impulses. One well-known aftereffect that is quite common is the afterhyperpolarization (AHP), which follows many action potentials and which indicates a refractory state (harder to elicit an action potential). There is evidence, in fact, that successive AHPs summate in spinal motoneurons (Kernell 1965). Another related aftereffect is the accumulation of potassium ions in the extracellular space, which leads to a shift in the resting potential and may possibly cause many complicated voltage-dependent conduction changes to take place.

Furthermore, the activity of a cell within a neuronal network may lead, via feedback paths, to self-excitation or self-inhibition and the amounts of these will depend on the frequency of firing.

If, however, a cell is in a spontaneously active state, firing at a relatively low mean rate, then the ISIs tend to be so long that there will be little interaction between neighboring ISIs. Notwithstanding the effects of trends in the input, the ISIs may then be approximately independent random variables.

In this section we briefly examine methods that have been employed to ascertain whether ISIs are correlated with those that follow or precede them.

(a) Autocorrelation coefficients of the ISIs

Autocorrelation coefficients have often been used to test for dependencies in trains of action potentials (Hagiwara 1954; Poggio and Viernstein 1964; Nakahama et al 1968; Muschawek and Loevner 1978).

In Section 9.2, Equation (9.16), we defined the covariance of two random variables. When this quantity is normalized by dividing by the product of the standard deviations of X and Y, one obtains the *correlation coefficient*

$$\rho_{XY} = \text{Cov}(X, Y)/\sigma_X\sigma_Y, \tag{10.75}$$

for which

$$-1 \le \rho_{XY} \le 1. \tag{10.76}$$

A random sample of size n for a bivariate population (X, Y) consists of the n pairs (X_k, Y_k), $k = 1, 2, \ldots, n$, and the *sample correlation coefficient* may be defined as the random variable

$$R_{XY} = \frac{\Sigma(X_k - \overline{X})(Y_k - \overline{Y})}{\sqrt{\Sigma_{k=1}^n(X_k - \overline{X})^2\Sigma_{k=1}^n(Y_k - \overline{Y})^2}} \tag{10.77}$$

where

$$\overline{X} = \frac{1}{n}\sum_{k=1}^n X_k,$$
$$\overline{Y} = \frac{1}{n}\sum_{k=1}^n Y_k \tag{10.78}$$

are the sample means for X and Y. Values of R_{XY} close to zero indicate that X and Y are uncorrelated, which sometimes implies they are independent.

In the analysis of a neuronal spike train, we are concerned with the sequence of random intervals $T_1, T_2, \ldots, T_n$. We consider pairs of consecutive variables (T_1, T_2), $(T_2, T_3), \ldots$ *as if* they were a sample from a bivariate population (X = an interval, Y = the next interval). Of course, the situation is now very different because the pairs (T_1, T_2) and (T_2, T_3), for example, have a member in common.

In the following it is assumed that each interval has the same distribution with mean μ and variance σ^2. We define the *serial correlation coefficient at lag* 1 or the *autocorrelation coefficient at lag* 1 by

$$\rho_1 = \frac{\text{Cov}(T_k, T_{k+1})}{\sigma^2}$$

$$= \frac{E((T_k - \mu)(T_{k+1} - \mu))}{\sigma^2}. \tag{10.79}$$

To estimate ρ_1 we use the random variable

$$R_1 = \frac{\sum_{k=1}^{n-1}(T_k - \bar{T}_*)(T_{k+1} - \bar{T}_{**})}{\sqrt{\sum_{k=1}^{n-1}(T_k - \bar{T}_*)^2 \sum_{k=1}^{n-1}(T_k - \bar{T}_{**})^2}}, \tag{10.80}$$

where

$$\bar{T}_* = \frac{1}{n-1} \sum_{k=1}^{n-1} T_k \tag{10.81}$$

is the sample mean for the first $n - 1$ intervals and

$$\bar{T}_{**} = \frac{1}{n-1} \sum_{k=1}^{n-1} T_{k+1} \tag{10.82}$$

is the sample mean for the last $n - 1$ intervals.

When n is large, (10.80) is usually replaced by the simpler expression

$$R_1 = \frac{\sum_{k=1}^{n-1}(T_k - \bar{T})(T_{k+1} - \bar{T})}{\sum_{k=1}^{n}(T_k - \bar{T})^2} \tag{10.83}$$

where $\bar{T}$ is the mean interval for the whole spike train.

It can be shown that if the T_k's are independent and n is large, then R_1 is approximately normally distributed with mean zero and variance $1/n$ [see, for example, Cox and Lewis (1966) or Chatfield (1975)]. Thus, as a rule of thumb, we may reject the null hypothesis of independence of neighboring ISIs whenever

$$|r_1| > 1.96/\sqrt{n}.$$

The probability of incorrectly rejecting H_0 when T_k and T_{k+1} are independent is then about 0.05.

Now, it is possible that consecutive intervals are independent but, for example, pairs of intervals "at lag 2" in the sequence, (T_1, T_3),

$(T_2, T_4), \ldots, (T_{n-2}, T_n)$ are not independent. In general, the *autocorrelation coefficient at lag j* is defined as

$$\rho_j = \mathrm{Cov}(T_k, T_{k+j})/\sigma^2. \tag{10.84}$$

An estimator for ρ_j is

$$R_j = \frac{\sum_{k=1}^{n-j}(T_k - \bar{T})(T_{k+j} - \bar{T})}{\sum_{k=1}^{n}(T_k - \bar{T})^2}. \tag{10.85}$$

Provided that $k \ll n$, R_j is, under the assumption that the T_k's are independent, also approximately normal with mean 0 and variance $1/n$.

It is useful to plot r_j versus j to obtain an *autocorrelogram* or *serial correlogram*. One may draw on the correlogram the $\pm 1.96/\sqrt{n}$ levels to see immediately which r_j's lie (individually) outside the acceptable limits for independence. An illustrative autocorrelogram with values of r_j up to $j = 7$ is shown in Figure 10.1 for a sequence of 100 intervals. It can be seen that r_1 and r_2 lie within the $\pm 1.96/\sqrt{n}$ limits, which makes it unlikely that T_k, T_{k+1} and T_k, T_{k+2} are correlated. However, r_3 is outside the limits and the evidence is strong that we should reject the claim that intervals three apart are independent. Note that r_6 is also outside the limits. In general, if r_j is large, then so too will be r_{jm} for $m = 2, 3, \ldots$. For sequences of neuronal interspike time intervals, it should not be necessary to examine r_j for j much larger than 10. Large values of r_j for j greater than about 10 are likely to be spurious. An extensive discourse on autocorrelation coefficients can be found in Cox and Lewis (1966).

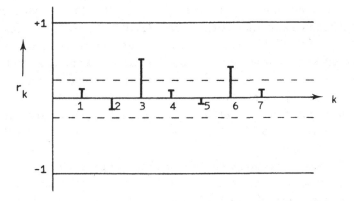

Figure 10.1. An autocorrelogram showing values of $r_1, \ldots, r_7$. Dashed lines are nominal 95% confidence limits, ± 0.196 with $n = 100$.

(b) *Joint density of consecutive ISIs*

For a *stationary sequence* of ISIs, as occurs in the ordinary renewal process, let the joint probability density of consecutive pairs of intervals be $f_{1,2}(t_1, t_2)$. Then, from the definition of joint density,

$$\Pr\{t_1 < T_k < t_1 + \Delta t_1, t_2 < T_{k+1} < t_2 + \Delta t_2\} \approx f_{1,2}(t_1, t_2) \, \Delta t_1 \, \Delta t_2.$$
$$(10.86)$$

It is useful as a preliminary step in the analysis of a spike train to plot the $n - 1$ pairs (t_k, t_{k+1}), $k = 1, \ldots, n - 1$, which gives insight into where in the positive quadrant the probability mass of the joint density is concentrated. This has been done for several kinds of neurons (Rodieck et al. 1962; O'Brien et al. 1973; Benoit and Chataignier 1973; Bassant 1976; Muschawek and Loevner 1978). An excess of points at large t_{k+1} and small t_k indicates a tendency for short intervals to be followed by long ones, and so forth.

If successive ISIs are independent, then the joint density $f_{1,2}$ factors into the product of two identical densities

$$f_{1,2}(t_1, t_2) = f(t_1)f(t_2). \tag{10.87}$$

Furthermore, if the random variables T_k and T_{k+1} are independent, we must have (see Section 9.2)

$$E(T_{k+1}|t_1 < T_k < t_1 + \Delta t_1) = E(T_{k+1}) = \mu, \tag{10.88}$$

and

$$E(T_k|t_2 < T_{k+1} < t_2 + \Delta t_2) = E(T_k) = \mu. \tag{10.89}$$

Rodieck et al. (1962) utilized these facts in a clever way to devise a test for independence. Taking the joint interval histogram as defined above, they compute (10.88) for various t_1 to obtain what they call "column means." Similarly, computing (10.89) for various t_2 results in "row means." If successive intervals are independent, the column means and the row means should all be approximately constant. Thus a plot of column mean versus t_1 should result in a constant μ, and so should a plot of row means against t_2. Yang and Chen (1978) have called this test the *conditional means test* and construed it as a standard one-way analysis of variance for testing the equality of several means (F-test).

We saw in Section 9.10.2 that the spectral density of a continuous-time wide-sense stationary process is the Fourier transform of its covariance kernel. For a wide-sense stationary discrete-time process $\{T_1, T_2, \ldots\}$, the analogous quantity is the discrete Fourier transform of the autocorrelation sequence ρ_j. This is called the *spectrum of intervals* and is defined as

$$f(\omega) = \frac{1}{2\pi} \sum_{j=-\infty}^{\infty} \rho_j e^{-ij\omega} = \frac{1}{2\pi} \sum_{j=-\infty}^{\infty} \rho_j (\cos j\omega - i \sin j\omega).$$

(10.90)

Since $f(\omega)$ so defined has period 2π in ω, we may restrict ω to the interval $[-\pi, \pi]$. Furthermore, since $\rho_j = -\rho_j$, contributions from $\sin(j\omega)$ drop out of (10.90) to leave

$$f(\omega) = \frac{1}{2\pi} \left[1 + 2 \sum_{j=1}^{\infty} \rho_j \cos(j\omega) \right].$$

(10.91)

Also, since it is clear that f is an even function of ω, we may further restrict ω to the interval $[0, \pi]$. If ρ_j is large only for $j = j^*$, then $f(\omega)$ will be dominated by an oscillatory component with period $2\pi/j^*$. If the intervals are independent and $\rho_j = 0$ for all j, then $f(\omega)$ will be constant (flat spectrum). Estimation of $f(\omega)$ is complicated as are tests based on it. The reader is advised to consult Bartlett (1963) and Cox and Lewis (1966). There are also several recent articles on spectral methods for analyzing neuronal spike-train data (Correia and Landolt 1977; Kwaadsteniet 1982).

10.8   Tests for Poisson processes
In 1952, Fatt and Katz made their celebrated Poisson hypothesis concerning spontaneous miniature endplate potentials (m.e.p.p.'s) at the frog neuromuscular junction. The histogram of time intervals between m.e.p.p.'s looked like it might be generated by an exponentially distributed random variable. A χ^2-test was used to support their hypothesis. Similar findings were reported later for slow muscle fibers of the frog (Burke 1957).

Subsequently, m.e.p.p. arrival times at various junctions have been scrutinized with a view to refuting or supporting the Poisson hypothesis (Cohen, Kita, and Van der Kloot 1973; Hubbard and Jones 1973; Cohen, Kita, and Van der Kloot 1974a, b; Van der Kloot et al. 1975; Washio and Inouye 1980). Similar methods have been employed to analyze neuronal ISIs [see, for example, Correia and Landolt (1977)] and it is interesting to note that early workers refer to a Poisson process as if it was the *only* random process. More recently, the possibility of a nonstationary Poisson process has been raised in connection with m.e.p.p.'s (Yana et al 1984).

Assume for now the process of interest is *stationary*, or *temporally homogeneous*. If $\{N(t), t \geq 0\}$ is a Poisson process, then the intervals between events (spikes or m.e.p.p.'s) must be (a) exponentially distributed and (b) independent. In fact, if both of these properties are demonstrated, then one may consider the process to be Poisson, within, of course, the bounds of statistical errors. Thus tests for a Poisson process have already been described, since Section 10.5 dealt with (a) and Section 10.7 dealt with (b). However, a number of statistical tests have been devised specifically to examine whether a random sequence of points is likely to have been generated by a stationary Poisson process. Two of these are the following.

(i) *The uniform conditional test*
This is based on the fact that if it is known an event has occurred in a Poisson process in the time interval $(0, s)$, then the time of occurrence is uniformly distributed on $(0, s)$. For details, see Cox and Lewis (1966), page 153.

(ii) *The C_α-test*
Let $\{N(t)\}$ be a Poisson process with intensity λ. One has therefore that $E(N(t)) = \lambda t = \text{Var}(N(t))$, exactly. By (10.15), since a Poisson process is a special case of a renewal process,

$$N(t) \xrightarrow[t \to \infty]{d} N(\lambda t, \sqrt{\lambda t}). \qquad (10.92)$$

The C_α-test examines the null hypothesis that $\{N(t)\}$ is a renewal process such that $N(t)$ is asymptotically normal with a mean and variance that are equal, versus the alternative hypothesis that $N(t)$ is a renewal process but that the asymptotic mean and variance of $N(t)$ are not equal. For details, see Yang and Chen (1978).

Nonstationary Poisson processes
In a temporally homogeneous, or stationary, Poisson process the rate parameter is constant in time. We now suppose that the rate may depend on time and put

$$\lambda(t) = \lim_{\Delta t \to 0} \frac{1}{\Delta t} \Pr\{\text{an event occurs in } (t, t + \Delta t]\}, \qquad (10.93)$$

to obtain a *nonstationary, or nonhomogeneous, Poisson process*. The remaining defining properties of a Poisson process are retained. In particular, the numbers of events in disjoint time intervals are independent.

It is not difficult to show that if $N(0) = 0$ and $N(t)$ is the number of events in the nonstationary process in $(0, t]$, then $N(t)$ is a Poisson random variable with mean

$$\Lambda(t) = \int_0^t \lambda(s)\, ds. \tag{10.94}$$

Some of the properties we are accustomed to enjoying for a stationary Poisson process no longer hold. In particular, time intervals between events are not independent and they are not identically distributed. To see this, let $T_1, T_2, \ldots$ be the first, second, $\ldots$ time intervals between events. Then

$$\Pr\{T_1 > t\} = \exp\left\{ - \int_0^t \lambda(s)\, ds \right\}, \tag{10.95}$$

so the density of T_1 is

$$f_{T_1}(t) = \lambda(t)\exp\left\{ - \int_0^t \lambda(s)\, ds \right\}, \qquad t > 0. \tag{10.96}$$

Given that $T_1 = t_1$, we find

$$\Pr(T_2 > t \mid T_1 = t_1) = \exp\left\{ - \int_{t_1}^{t_1 + t} \lambda(s)\, ds \right\}. \tag{10.97}$$

It is left as an exercise to show that this leads to the conditional density for T_2,

$$f_{T_2}(t \mid t_1) = \lambda(t_1 + t)\exp\left\{ - \int_{t_1}^{t_1 + t} \lambda(s)\, ds \right\}, \qquad t_1, t > 0, \tag{10.98}$$

and that unconditionally we have

$$\Pr(T_2 > t) = \int_0^\infty \exp\left\{ - \int_0^{t_1 + t} \lambda(s)\, ds \right\} \lambda(t_1)\, dt_1. \tag{10.99}$$

The analysis of such processes is facilitated by the fact that they may be transformed to stationary Poisson processes by means of a change of time scale. One sees that if

$$\tau = \int_0^t \lambda(s)\, ds,$$
$$J(\tau) = N(t), \tag{10.100}$$

then $\{ J(\tau), \tau \geq 0 \}$ is a stationary Poisson process such that the number of events in $(\tau_1, \tau_2]$ has the law

$$\Pr(J(\tau_1, \tau_2) = k) = \frac{(\tau_2 - \tau_1)^k e^{-(\tau_2 - \tau_1)}}{k!}, \qquad k = 0, 1, 2, \ldots .$$

(10.101)

Thus J has unit intensity. Furthermore, the random variables

$$S_k = \int_{T_{k-1}}^{T_k} \lambda(s)\, ds, \qquad k = 1, 2, \ldots,$$

(10.102)

with $T_0 \doteq 0$, are independent with common density e^{-t}.

Of course, before any analysis can be performed on the transformed process, one must estimate the rate function $\{\lambda(t),\ t \geq 0\}$. One method is to maximize a likelihood function. A suitable one given by Cox and Hinkley (1975) for an observation period $(0, t_0]$ is

$$L(\{\lambda(t), t \in (0, t_0]\}; t_1, \ldots, t_n) = \exp\left\{ -\int_0^{t_0} \lambda(s)\, ds \right\} \prod_{k=1}^{n} \lambda(t_k),$$

(10.103)

where $t_1, t_2, \ldots, t_n$ are the times of occurrence of events. This has been utilized by Yana et al. (1984) in their analysis of m.e.p.p.'s at the frog neuromuscular junction. A popular choice for $\lambda(t)$ (Cox and Lewis 1966) has been

$$\lambda(t) = \alpha e^{\beta t},$$

(10.104)

where $\alpha > 0$. Then the logarithm of the above likelihood function is

$$\ln L = \frac{\alpha}{\beta}(1 - e^{\beta t_0}) + \beta \sum_{k=1}^{n} t_k + n \ln \alpha.$$

(10.105)

Setting $\partial \ln L / \partial \alpha$ and $\partial \ln L / \partial \beta$ to zero, we obtain the following equations that must be satisfied by the maximum-likelihood estimates of α and β,

$$\sum_{k=1}^{n} t_k + n \left(\frac{1}{\hat{\beta}} + \frac{t_0}{e^{-\hat{\beta} t_0} - 1} \right) = 0,$$

(10.106)

$$\hat{\alpha} = \frac{n\hat{\beta}}{e^{\hat{\beta} t_0} - 1}.$$

(10.107)

The first of this pair of equations is solved numerically for $\hat{\beta}$, whereupon $\hat{\alpha}$ may be found from the second equation. The battery of available tests can then be performed on the process in the transformed time scale. This procedure was carried out by Yana et al. (1984).

10.9   Tests for a renewal process

A renewal process is specified in terms of the independent and identically distributed intervals $\{T_1, T_2, \ldots\}$. Tests for a renewal process must therefore consider the aspects of independence and the common probability distribution. One of the chief complicating factors is stationarity, since nonstationarity may mask independence.

Tests for independence have already been discussed in Section 10.7. The use of autocorrelation coefficients will not usually be hampered by the lack of knowledge of their distributions for small samples, since large samples are readily obtained. The conditional means test of Rodieck et al. (1962) is useful but there is no version of the test that takes into account the possibility of nonstationarity.

Yang and Chen (1978) utilize the asymptotic normality of $N(t)$ as $t \to \infty$ as in Equation (10.15) to test the null hypothesis of a renewal process against the alternative of a nonrenewal process. Essentially, the test ascertains whether the right conditions prevail for the application of the central limit theorem.

To execute the test, a spike train is subdivided into m subtrains of duration t_1 and the number of spikes N_i in subtrain i obtained. If t_1 is large and the renewal hypothesis is true, the $\{N_i\}$ will be a sample from a normal distribution with mean t_1/μ and variance $\sigma^2 t_1/\mu^3$.

Equating moments or maximizing likelihood gives the estimates

$$\hat{\mu} = \frac{t_1}{\overline{N}}, \tag{10.108}$$

$$\hat{\sigma}^2 = \frac{t_1^2}{(m-1)\overline{N}^3} \sum_{i=1}^{m} (N_i - \overline{N})^2, \tag{10.109}$$

where

$$\overline{N} = \frac{1}{m} \sum_{i=1}^{m} N_i$$

is the mean number of spikes in an interval of length t_1. One may then test whether the $\{N_i\}$ are likely to have been generated by an $N(\hat{\mu}, \hat{\sigma})$ distribution using either a χ^2-test or a Kolmogorov–Smirnov test (see Section 10.5).

There have not been many tests performed on spike trains for the renewal hypothesis although many authors have tested for independence. One extensive study was that of Correia and Landolt (1977) on hair-cell receptors of the pigeon vestibular apparatus. Using several diverse tests, these authors found lack of evidence contradicting the renewal model in 64% of the cells studied. Ekholm and Hyvärinen

(1970) reported in another study that 70% of the 300 cells satisfied the renewal hypothesis.

10.10 Generalizations and modifications to the standard renewal model

The basic renewal process, consisting essentially of a sequence of independent and identically distributed random variables, is just one class of stationary point processes. By virtue of the independence assumption, the serial correlation coefficients of all lags must be zero for a standard renewal process. Some modifications and generalizations of this basic model have been proposed and are discussed briefly in this section. The underlying idea is that a cell may be in one of several "states," each state being characterized by a different ISI distribution. One further supposes that the neuron changes from state to state in possibly a random fashion.

In this section we will begin with a two-state semi-Markov process model, which is actually a special case of the model considered afterward. This makes more obvious the relation between the special case and the general model. Finally, we consider a related model due to Smith and Smith (1965). For a comprehensive treatment of stationary point processes, the reader is advised to consult Cox and Lewis (1966) and Snyder (1975).

(i) *Two-state semi-Markov process*

It is assumed that there are two neuronal states. When a neuron is in state 1 the ISI has density f_1, whereas in state 2 the ISI density is f_2. Let the sequence of states at epochs $0, 1, 2, \ldots$ be $\{X_0, X_1, X_2, \ldots\}$. A spike occurs at the beginning of the experiment and the density of the first ISI is chosen. Subsequently, if the cell goes into state i at epoch n, then the $(n+1)$st ISI $(= T_{n+1})$ is drawn from the density f_i. We suppose that the sequence of states is a Markov chain with transition matrix

$$\mathbf{P} = \begin{bmatrix} p_{11} & p_{12} \\ p_{21} & p_{22} \end{bmatrix}. \tag{10.110}$$

Here the p_{jk}'s are the transition probabilities that give the conditional probability of state k's being occupied at epoch $n+1$ given that state j was occupied at epoch n,

$$\Pr(X_{n+1} = k | X_n = j) = p_{jk}, \qquad j, k = 1, 2; n = 0, 1, 2, \ldots. \tag{10.111}$$

Since from state 1 transitions must be to state 1 or 2, we must have

$p_{11} + p_{12} = 1$ and $p_{21} + p_{22} = 1$. Hence if we put $p_{11} = \alpha_1$ and $p_{22} = \alpha_2$, the above matrix may be simplified to

$$\mathbf{P} = \begin{bmatrix} \alpha_1 & 1 - \alpha_1 \\ 1 - \alpha_2 & \alpha_2 \end{bmatrix}. \tag{10.112}$$

Define

$$\pi_1 = \frac{1 - \alpha_2}{2 - \alpha_1 - \alpha_2}, \tag{10.113}$$

$$\pi_2 = \frac{1 - \alpha_1}{2 - \alpha_1 - \alpha_2}, \qquad \alpha_1 + \alpha_2 \neq 2; \tag{10.114}$$

then it may be shown that, providing not both α_1 and α_2 are unity, the mean and variance of the ISI are

$$E(T) = \pi_1 \mu_1 + \pi_2 \mu_2, \tag{10.115}$$

$$\text{Var}(T) = \pi_1 \sigma_1^2 + \pi_2 \sigma_2^2 + \pi_1 \pi_2 (\mu_1 - \mu_2)^2, \tag{10.116}$$

where μ_i and σ_i^2 are the mean and variance for f_i. Furthermore, the *density of T* is

$$f(t) = \pi_1 f_1(t) + \pi_2 f_2(t), \tag{10.117}$$

and the serial correlation coefficient at lag k is

$$\rho_k = \frac{(\mu_1 - \mu_2)^2 \pi_1 \pi_2 \beta^k}{\pi_1 \sigma_1^2 + \pi_2 \sigma_2^2 + \pi_1 \pi_2 (\mu_1 - \mu_2)^2}, \qquad k = 1, 2, \ldots, \tag{10.118}$$

where

$$\beta = \alpha_1 + \alpha_2 - 1. \tag{10.119}$$

The following special cases are of interest.

(a) $\alpha_1 = \alpha_2 = 1$.

$$\mathbf{P} = \begin{bmatrix} 1 & 0 \\ 0 & 1 \end{bmatrix}. \tag{10.120}$$

There are no changes of state. If the cell starts in state 1, it remains in state 1 so all intervals are drawn from the density f_1. This gives the *original renewal model*.

(b) $\alpha_1 = 0, \alpha_2 = 1$ or $\alpha_1 = 1, \alpha_2 = 0$. Consider the first of these cases with

$$\mathbf{P} = \begin{bmatrix} 0 & 1 \\ 0 & 1 \end{bmatrix}. \tag{10.121}$$

If the cell starts in state 1, it goes to state 2 with probability 1 and must stay there. If it starts in state 2, there are no intervals of type 1. Hence this is the renewal model with a possibly different first-interval distribution; that is the *modified* renewal process.

(c) $\alpha_1 = \alpha_2 = 0$. Then

$$\mathbf{P} = \begin{bmatrix} 0 & 1 \\ 1 & 0 \end{bmatrix}. \tag{10.122}$$

At each epoch a change of state must occur. Hence intervals are alternately drawn from f_1 and f_2. This is called an *alternating renewal process* and the ISI will have density $f = (f_1 + f_2)/2$.

(d) $\alpha_1 + \alpha_2 = 1$; $\alpha_1, \alpha_2 \neq 0$. We may put $\alpha_1 = \alpha$ and $\alpha_2 = 1 - \alpha$ with $0 < \alpha < 1$. We have

$$\mathbf{P} = \begin{bmatrix} \alpha & 1 - \alpha \\ \alpha & 1 - \alpha \end{bmatrix}. \tag{10.123}$$

There is always a nonzero chance that a change of state will occur at any epoch. Clearly, the long-run fraction of intervals of type 1 is α, whereas for type 2 this quantity is $1 - \alpha$. The process is a renewal process and the density of the ISI is (using the law of total probability)

$$f(t) = \alpha f_1(t) + (1 - \alpha) f_2(t). \tag{10.124}$$

A special case of interest is that in which both f_1 and f_2 are exponential with means $1/\lambda_1$ and $1/\lambda_2$. Then

$$f(t) = \alpha \lambda_1 e^{-\lambda_1 t} + (1 - \alpha)\lambda_2 e^{-\lambda_2 t}, \tag{10.125}$$

with

$$E(T) = \frac{\alpha}{\lambda_1} + \frac{1 - \alpha}{\lambda_2}, \tag{10.126}$$

$$\mathrm{Var}(T) = \frac{\alpha(2 - \alpha)}{\lambda_1^2} + \frac{1 - \alpha^2}{\lambda_2^2} - \frac{2\alpha(1 - \alpha)}{\lambda_1 \lambda_2}. \tag{10.127}$$

The coefficient of variation of the ISI is

$$\mathrm{CV}(T) = \sqrt{\mathrm{Var}(T)}/E(T), \tag{10.128}$$

which is greater than 1 for all $\lambda_1 \neq \lambda_2$ and $0 < \alpha < 1$. The proof of this is left as an exercise. Parameter estimation for densities such as (10.125) by analytical methods seems difficult. It will also be found that the number of ISIs of a given type has a geometric distribution. Thus if N_1 is the number of consecutive intervals of type 1,

$$\Pr(N_1 = m) = (1 - \alpha)\alpha^{m-1}, \qquad m = 1, 2, \ldots, \tag{10.129}$$

since a run of length m must terminate with an interval of type 2. Similarly for N_2, the number of consecutive intervals of type 2.

(ii) *Randomly alternating renewal process*

In the two-state semi-Markov process model, the number of consecutive ISIs drawn from f_1 or f_2 is a geometrically distributed random variable. The numbers of intervals so drawn may have other distributions and this leads to the model we will discuss. It was introduced by Ekholm, elaborated on and applied by Ekholm and Hyvärinen (1970), and subsequently employed and further investigated by Kwaadsteniet (1982). The model is called *pseudo-Markov* by Ekholm and Hyvärinen and *semialternating renewal* by Kwaadsteniet.

In the present model there are also two possible neuronal states with corresponding ISI densities f_1 and f_2. With these states there are associated positive integer-valued random variables N_1 and N_2 with probability mass functions $p_{1,k}$ and $p_{2,k}$, respectively. That is,

$$\Pr(N_i = k) = p_{i,k}; \qquad i = 1,2; \; k = 1,2,\ldots . \tag{10.130}$$

Let $N_1^{(k)}$ and $N_2^{(k)}$, $k = 1,2,\ldots$, be random samples for N_1 and N_2. Suppose the cell is initially in state 1. Then $N_1^{(1)}$ ISIs are drawn from the density f_1; these are followed by $N_2^{(1)}$ ISIs from the density f_2; then come $N_1^{(2)}$ ISIs from f_1, and so forth. Thus the ISI distribution alternates back and forth between those of f_1 and f_2 and the number of spikes emitted in each state is random: hence the name *randomly alternating renewal process*.

The structure of the process may be exhibited succinctly by letting $T_1^{(k)}$ and $T_2^{(k)}$, $k = 1,2,\ldots$, be random samples for variables with densities f_1 and f_2, respectively. Define a run as a sequence of ISIs from the same population. Let the first ISI be drawn from f_1. The first run of ISIs is then of duration

$$S_1 = T_1^{(1)} + \cdots + T_1^{N_1^{(1)}}; \tag{10.131}$$

the second has duration

$$S_2 = T_2^{(1)} + \cdots + T_2^{N_2^{(1)}}. \tag{10.132}$$

In general,

$$S_{2k+1} = T_1^{N_1^{(1)} + \cdots + (N_1^{(k)} + 1)} + \cdots + T_1^{N_1^{(1)} + \cdots + N_1^{(k+1)}}, \qquad k = 0,1,\ldots, \tag{10.133}$$

$$S_{2k} = T_2^{N_2^{(1)} + \cdots + (N_2^{(k-1)} + 1)} + \cdots + T_2^{N_2^{(1)} + \cdots + N_2^{(k)}}, \qquad k = 1,2,\ldots . \tag{10.134}$$

Each S_n is therefore a compound random variable.

This model contains the ordinary renewal process and two-state semi-Markov process models as special cases. The mean and variance of the ISI and the correlation coefficients for the ISI sequence have been obtained by Ekholm and Hyvärinen (1970). An application to spike trains in neurons of the pond snail can be found in Kwaadsteniet (1982).

(iii) *Randomly blocked Poisson process*

Smith and Smith (1965) introduced the following model to account for the nature of the spontaneous activity of cat *isolated cortical neurons*. Many of those cells undergo periods of quiet separated by periods of bursting activity. Again there are two basic states for the cell, but spikes do not coincide with every change of state. A description of the model is as follows:

(a) The cell alternates between two states, which we shall call active and resting. The time spent in the active state is exponentially distributed with mean $1/\lambda$ and the time spent in the resting state is similarly distributed but with mean $1/\mu$. The time intervals spent in either state are independent.
(b) At the beginning of each active period a spike is emitted and this switches on a Poisson process of intensity ν. Spikes occur at the event times in this Poisson process. Events in the Poisson process occur independently of the time intervals spent in the resting and active states.
(c) There are no spikes while the cell is in the resting state.

Another way to construe this method of generation of an ISI sequence is the following. Spikes occur at event times in a Poisson process of rate ν, called the *background* Poisson process. This background process is blocked for periods that are exponentially distributed with mean $1/\mu$ and the periods of blocking are separated by periods that are exponentially distributed with mean $1/\lambda$. A spike always accompanies a transition from the blocked to unblocked state but not on the reverse transitions.

The ISI density

The ISI density for the present model is much easier to find than it would be without the assumption that a resting period terminates with a spike. We note that there are only two ways in which an ISI (random variable T) may fall in the interval $(t, t + \Delta t]$, assuming, of course, a spike occurred at $t = 0$. These are:

A. The active state persisted throughout $(0, t]$ and the first event occurred in the background Poisson process in $(t, t + \Delta t]$. The probability of these contingencies is, to $o(\Delta t)$,

$$f_{(1)}(t) \, \Delta t = e^{-\lambda t} \nu e^{-\nu t} \Delta t. \tag{10.135}$$

B. The active state persisted until some time $t' < t$, there being no event in the background Poisson process in $(0, t']$, and then the resting state persisted from t' to t and concluded with a spike. Integrating over all possible values of t' gives

$$f_{(2)}(t) \, \Delta t = \int_0^t e^{-\nu t'} \lambda e^{-\lambda t'} \, dt' \mu e^{-\mu(t - t')} \Delta t$$

$$= \frac{\lambda \mu e^{-\mu t}}{\nu + \lambda - \mu} (1 - e^{-t(\nu + \lambda - \mu)}) \, \Delta t. \tag{10.136}$$

Since A and B are mutually exclusive and their union is the sample space, we have

$$f_T(t) = f_{(1)}(t) + f_{(2)}(t)$$

$$= \nu e^{-(\nu + \lambda)t} + \frac{\lambda \mu}{\nu + \lambda - \mu} (e^{-\mu t} - e^{-(\nu + \lambda)t}), \tag{10.137}$$

a result obtained somewhat less directly by Smith and Smith (1965). This may be put in the form

$$f_T(t) = \alpha(\nu + \lambda) e^{-(\nu + \lambda)t} + (1 - \alpha)\mu e^{-\mu t}, \tag{10.138}$$

which is the same as (10.125). Hence the mean and variance of T may be obtained from (10.126) and (10.127) by putting $\nu + \lambda$ for λ_1, μ for λ_2, and

$$\alpha = \frac{\nu - \mu}{\nu + \lambda - \mu}, \tag{10.139}$$

$$1 - \alpha = \frac{\lambda}{\nu + \lambda - \mu}. \tag{10.140}$$

As noted above, the coefficient of variation of the ISI must be greater than one. Smith and Smith found that each of the 40 cells they examined had ISI histograms that were well fitted by densities of the form (10.138).

The picture emerged that there were two processes, a fast and a slow, operating in these cells. The slower process determines the onset times of bursts of action potentials, whereas the fast process governs the time intervals between spikes within a burst. Interestingly, it was found that the application of electric current could change the frequency of the bursts, but the fast process was insensitive to such stimulation. It was concluded that the fast process is somehow governed by factors intrinsic to the cell.

10.11 Parameter estimation for diffusion processes representing nerve membrane potential

In Section 10.3 methods were described for estimating parameters in prescribed probability density functions (first-passage time densities, in particular) in order to obtain good fits to ISI histograms. The ultimate aim of such procedures is to infer physiological and anatomical properties from ISI data. As such, this has proven to be a difficult and challenging problem. One of its motivating factors has been that small cells in the CNS are difficult to record from intracellularly but ISIs may be recorded extracellularly. In the event that intracellular recording is possible, there is still useful parameter estimation to be done on nerve membrane-potential depolarization as pointed out by Lansky (1983).

We will use two approaches to maximum-likelihood estimation of parameters for diffusion processes. One of these is that used in Section 10.3, which we will call *estimation from the probability density function*. The second method is based on having a continuous record of the diffusion over a given time interval and we will call this *estimation from sample paths*. The processes we will consider are the Wiener process with drift (Section 9.6) and the Ornstein–Uhlenbeck process (Section 9.9). We will examine the second method first.

(i) *Estimation of parameters from sample paths*

The theory of this method of *maximum-likelihood estimation* of parameters for diffusion processes involves martingales and Radon–Nikodym derivatives and will not be given here. However, Feigin (1976) has made some general results quite accessible.

Consider the diffusion process $\{X(t), t \geq 0\}$ whose Itô differential (see Section 9.7) is

$$dX = \mu(X)\, dt + \sigma(X)\, dW, \qquad (10.141)$$

and assume $X(0)$ is given.

Partition $[0, s]$ into 2^n equal subintervals of length $2^{-n}s$ and let

$$t_j = 2^{-n}js, \qquad j = 0, 1, \ldots, 2^n.$$

It is known that for X satisfying (10.141),

$$\lim_{n \to \infty} \sum_{j=1}^{2^n} \left(X(t_{j+1}) - X(t_j) \right)^2 = \int_0^s \sigma^2(X(t))\, dt, \quad \text{a.s.,}$$

(10.142)

The quantity on the left-hand side is called the *quadratic variation process* of X. (The letters a.s. mean almost sure convergence; i.e., with probability 1.)

For the processes of interest to us here, the infinitesimal variance $\sigma^2(X)$ is constant at σ^2, say. Hence the right-hand side of (10.142) is just $\sigma^2 s$ and the left-hand side may be obtained approximately from a sample path by using the observed values of $X(t_1), X(t_2), \ldots, X(t_{2^n})$ for large n. Thus we may write

$$\hat{\sigma}^2 = \frac{1}{s} \int_0^s \left(dX(t) \right)^2.$$

(10.143)

We may consider therefore that σ has been estimated and that only parameters in the drift remain to be found.

Put $\mu = \mu(X(t); \theta)$ where θ is a parameter. Then the maximum-likelihood estimate $\hat{\theta}$ of θ is found by solving

$$\int_0^s \left[\frac{\frac{\partial \mu}{\partial \theta}(X(t); \theta)}{\sigma^2(X(t))} \right] dX(t) = \int_0^s \left[\frac{\mu(X(t); \theta) \frac{\partial \mu}{\partial \theta}(X(t); \theta)}{\sigma^2(X(t))} \right] dt.$$

(10.144)

We now apply this to two diffusions.

(a) *Wiener process with drift.* Here

$$dX = \mu\, dt + \sigma\, dW,$$

(10.145)

where μ is the only parameter to be estimated. Substituting in (10.144), we have

$$\int_0^s dX(t) = \int_0^s \mu\, dt,$$

so

$$X(s) - X(0) = \mu s,$$

or

$$\hat{\mu} = \frac{X(s) - X(0)}{s}.$$

(10.146)

Note the advantage of this method in that one does not need to know the transition density function; only the form of the infinitesimal moments.

(b) *Ornstein–Uhlenbeck process.* Here there are two parameters in the drift when we put

$$dX = (\mu - \beta X)\, dt + \sigma\, dW. \tag{10.147}$$

To estimate μ we assume β is known and vice versa. Applying (10.144), we obtain

$$\int_0^s dX(t) = \hat{\mu}s - \hat{\beta}\int_0^s X(t)\, dt,$$

$$\int_0^s X(t)\, dX(t) = \hat{\mu}\int_0^s X(t)\, dt - \hat{\beta}\int_0^s X^2(t)\, dt. \tag{10.148}$$

This system of equations may be solved for $\hat{\mu}$ and $\hat{\beta}$. Using the formula

$$\int_0^s X(t)\, dX(t) = \frac{1}{2}\left(X^2(s) - X^2(0) - s\right),$$

which follows from Itô's rule [see, for example, Liptser and Shiryayev (1977)], we finally obtain

$$\hat{\mu} = \frac{(X(s) - X(0))\int_0^s X^2(t)\, dt - \frac{1}{2}(X^2(s) - X^2(0) - s)\int_0^s X(t)\, dt}{s\int_0^s X^2(t)\, dt - \left(\int_0^s X(t)\, dt\right)^2},$$

$$\hat{\beta} = \frac{(X(s) - X(0))\int_0^s X(t)\, dt - \frac{s}{2}(X^2(s) - X^2(0) - s)}{s\int_0^s X^2(t)\, dt - \left(\int_0^s X(t)\, dt\right)^2}. \tag{10.149}$$

(ii) *Estimation of parameters from the transition probability density function*

(a) *Wiener process with drift.* We have, from Section 9.6, that the transition probability density function for the process (10.145) is

$$f(x, t; x_0, \mu, \sigma^2) = \frac{1}{\sqrt{2\pi\sigma^2 t}} \exp\left\{ -\frac{(x - x_0 - \mu t)^2}{2\sigma^2 t} \right\}. \tag{10.150}$$

Let the observed values of the depolarization be

$$X(t_k) = x_k, \qquad k = 1, 2, \ldots, n, \tag{10.151}$$

and assume for simplicity that $x_0 = 0$. Then we define a likelihood function

$$L = \prod_{k=1}^{n} \left(2\pi\sigma^2 t_k\right)^{-1/2} \exp\left\{ -\frac{(x_k - \mu t_k)^2}{2\sigma^2 t_k} \right\}. \tag{10.152}$$

Taking logarithms and solving the equations

$$\frac{\partial \ln L}{\partial \mu} = \frac{\partial \ln L}{\partial \sigma^2} = 0, \tag{10.153}$$

we obtain the following maximum-likelihood estimates for μ and σ^2:

$$\hat{\mu} = \left(\sum_{k=1}^{n} x_k \right) \bigg/ \left(\sum_{k=1}^{n} t_k \right),$$

$$\hat{\sigma}^2 = \frac{1}{n} \sum_{k=1}^{n} \left(\frac{x_k - \hat{\mu} t_k}{\sqrt{t_k}} \right)^2. \tag{10.154}$$

(b) *Ornstein–Uhlenbeck process.* The transition density for a process satisfying (10.147) with $\beta = 1$ and $X(0) = x_0$ is (see Section 9.9)

$$f\left(x, t; x_0, \mu, \sigma^2\right)$$

$$= \frac{1}{\sqrt{\pi\sigma^2(1 - e^{-2t})}} \exp\left\{ -\frac{\left(x - x_0 e^{-t} - \mu(1 - e^{-t})\right)^2}{\sigma^2(1 - e^{-2t})} \right\}. \tag{10.155}$$

For simplicity, we put $x_0 = 0$ and suppose that there are available the n values $X(t_k) = x_k$ of the subthreshold membrane potential. The problem of estimating μ and σ^2 is formally the same as the Wiener process with drift. One finds after some algebra that the maximum-likelihood estimates for μ and σ^2 are

$$\hat{\mu} = \left(\sum_{k=1}^{n} x_k \right) \bigg/ \sum_{k=1}^{n} (1 - e^{-2t_k}), \tag{10.156}$$

and

$$\hat{\sigma}^2 = \frac{2}{n} \sum_{k=1}^{n} \left(\frac{x_k - \hat{\mu}(1 - e^{-t_k})}{\sqrt{1 - e^{-t_k}}} \right)^2. \tag{10.157}$$

10.12   The classification and interpretation of ISI distributions

Examination of histograms of ISIs for several kinds of neuron in various animals enables us to distinguish broadly 10 types of

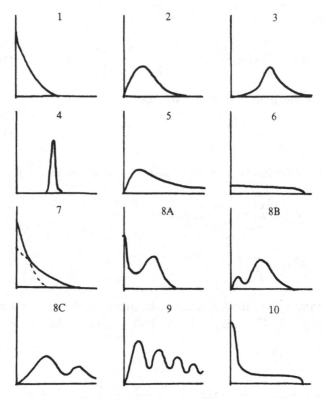

Figure 10.2. The 10 basic kinds of ISI distributions.

density functions although there is some overlap. These types are labeled 1–7, 8A, 8B, 8C, and 9 and are sketched in Figure 10.2.

Unimodal densities

Type 1.   The type-1 density is *exponential*, which is not infrequently encountered. In the case of neuronal ISIs one may conclude that an exponential distribution implies very large EPSPs were randomly arriving and being transformed on a one-to-one basis to action potentials. This is based, of course, on the fact that the waiting time for one event in a simple Poisson process has an exponential distribution.

Type 2.   The type-2 density is *low-order gammalike*. It is the most frequently encountered ISI density. Based on the results of simulations and the few known exact results for first-passage time densities for Markov processes, one would suspect that such an ISI distribution signifies that only a few EPSPs occurring in a short enough time interval are sufficient to make the neuron fire. Furthermore, it is unlikely that the cell is receiving significant quantities of inhibition. Sometimes lognormal distributions may be fitted to such ISI distributions.

Type 3. The type-3 density is *normal or Gaussian-like* and, of course, there is overlap with type 2. One would suspect that a large number of EPSPs are needed to make the cell fire and hence the threshold is large relative to the average EPSP amplitude. It is unlikely that significant inhibition is present.

Type 4. The type-4 density is *deltalike*, (i.e., it resembles a point mass). Of course, this is also the shape of a Gaussian density with a very small standard deviation. There is very little variability about the mean interval and the firing is almost deterministic. It is the kind of ISI distribution one expects from a pacemaker cell. Thus the cell is firing regularly by virtue of intrinsic mechanisms and there is some noise present that produces some scatter about the mean. The ISI distribution in case 9 is also of this shape, but here the cell was being driven by an external periodic stimulus that made it fire at regular intervals.

Type 5. Type-5 densities are characterized as *long-tailed densities*. These may occur in conjunction with types 2 and 3 in the sense that they resemble low-order gamma or normal densities at small t. A possible explanation of the long tails of these densities is that the cells are receiving significant amounts of inhibition. The densities are nevertheless unimodal.

Uniform distribution: *type* 6

The sixth type of ISI distribution resembles a uniform distribution. It is hard to imagine a single random mechanism of the kind that might operate in nerve cells to produce an approximately uniform distribution of ISIs. One possibility is that the cell exists in several states, either intrinsically or relative to its input patterns, and hence the ISI density is a linear combination of several densities with different means so that their sum is a flattened out uniform-looking density.

Bimodal densities

Type 7. This type is not bimodal in the sense of having two peaks but it is a *linear combination of two exponential densities*. The mechanisms for producing such a density, involving a change of state, have been dealt with in Section 10.10.

Types 8A, 8B, *and* 8C. These are all bona fide *bimodal densities*. In type 8A the additional peak occurs very close to the origin. It is apparent that the peak near the origin is a manifestation of a cell's firing in doublets or triplets, and so on, of action potentials occurring in groups very close together in time. Such patterns are sometimes called bursts. Renshaw cells, for example, can emit a burst of 10 or more spikes at a frequency of about 1000 impulses/s.

In type 8B one of the modes is close to the origin but not close enough to indicate bursting although this cannot be excluded. In type 8C the smaller modal ISI is quite remote from the origin therefore ruling out the idea of bursts. The only possible explanation in the case of spontaneous activity seems to be a change of state so that the ISI density is a linear combination of ISIs for the two states. If the larger mode is exactly twice the shorter mode, one may conclude that sometimes an action potential is being missed. Hence if T is an ISI, then the ISI distribution is a linear combination of the density of T and of $2T$.

Multimodal distributions: *type* 9. The density is called *multimodal* if the number of peaks exceeds 2. Such distributions are found for cells in the cat cochlear nucleus, when the animal was subjected to clicks at regular intervals. Hence the multimodality arises because sometimes one, two, three, etc., clicks did not give rise to a spike. The density is a linear combination of the densities of T, $2T$, $3T$, etc., where T is some random variable with a sharply peaked Gaussian-like density. The same remarks apply to some visual cells and indeed the explanation given by the experimentalists (Bishop et al. 1964) was that inhibitory inputs occasionally blocked the excitatory ones that were causing the lateral geniculate neuron to fire.

L-*shaped*: *type* 10. This is really a combination of types 6 and 8A. That is, the density is essentially uniform except for a sharp peak at the origin that makes the density L-shaped. As a final remark, if a cell only requires one excitatory input to fire, then the time between inputs will have about the same density as the ISI. This could happen for *any* of the above types.

10.13 Post-stimulus time histograms

Poststimulus time (PST) histograms are obtained in stimulus–response experiments on single neurons. Usually a sensory pathway (e.g., auditory, tactile, visual) is involved and the stimulus is a physical one corresponding to that sense modality. Sometimes, however, the stimulus may be an electric shock delivered to some connecting part of the nervous system. The time at which the stimulus

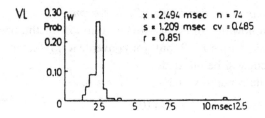

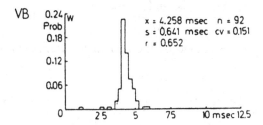

Figure 10.3. Poststimulus time histograms of a PT cell in response to electrical stimulations of the ventrolateral (upper histogram) and ventrobasal thalamic nuclei. [Adapted from Steriade et al. (1973).]

is delivered is known to the experimenter. Records are made of the spiking activity of a target neuron, which may be one or several levels of neurons away in the sensory pathway. Examples may be found in the works of Gerstein and Kiang (1960), Steriade et al. (1973), Gebber (1975), Whitsel et al. (1977), and Ryan and Miller (1977) this being just a small sample.

Two PST histograms are shown in Figure 10.3 for a pyramidal tract (PT) cell of the cat motor cortex [from Steriade et al. (1973)]. The stimulus in the top PST histogram was an electric shock (of several volts for less than one-tenth of a millisecond) delivered to the ventro-lateral thalamic nucleus. In the lower PST histogram, the shock was delivered to the ventrobasal thalamic nucleus.

Let us now make *three simplifying assumptions*:

(i) The cell whose responses are being observed does not emit spikes in the absence of the deliberate experimental stimulation.

(ii) The stimulus does, in fact, by some chain of events, cause the cell under observation to fire.

(iii) The time interval between stimulus presentations is sufficiently large so that there is no interference between the responses to successive stimuli.

Definitions in relation to the PST histogram

In the following, all events are conditioned on the presentation of a stimulus at time $t = 0$, but, for convenience, reference to the conditioning event will be omitted.

First, define the random variables $N(t)$, $t \geq 0$, as

$$N(t) = \text{number of spikes in } (0, t]. \tag{10.158}$$

That is, $\{ N(t), t \geq 0 \}$ is a counting process that jumps by one at the random times of occurrence of spikes. We may also define a *spike density* (cf. Section 10.2).

$$u_s(t) = \lim_{\Delta t \to 0} \frac{\Pr\{\text{a spike in } (t, t + \Delta t]\}}{\Delta t}, \qquad t > 0, \tag{10.159}$$

which is the quantity estimated by the *PST histogram*. As in Section 10.2, we also have

$$u_s(t) = \frac{d}{dt} E(N(t)). \tag{10.160}$$

There is another quantity of interest, namely, the appearance time of the *first spike* after the stimulus is presented. Define the random variable

$$T_L = \begin{cases} \inf\{t \mid N(t) \geq 1\}, \\ \infty \quad \text{if the above set is empty.} \end{cases} \tag{10.161}$$

Then T_L is the *latency* of the response.

Since in most, if not all cases, it is safe to assume a nonzero chance of no response at all, we will often have

$$\Pr\{T_L < \infty\} < 1, \tag{10.162}$$

in which case T_L will have neither a finite mean nor finite higher-order moments.

It is not generally possible to obtain an estimate of the probability density of T_L from the PST histogram. The PST histogram and the histogram of conditional first-spike times should both be reported. However, it is possible to estimate the minimum value of T_L, as this must coincide with the smallest value of t for which $N(t) > 0$.

In Figure 10.3, with reference to the top PST histogram, the minimum latency of the response is about 1.8 ms, whereas for the bottom PST histogram, the minimum latency is about 3 ms, if one ignores the seemingly spurious earlier occurring spikes. It was deduced that the pathway from the ventrolateral nucleus to the PT cell is a direct monosynaptic one. On the other hand, the pathway from the ventrobasal nucleus has perhaps one or two interneurons in it, making a di- or possibly trisynaptic connection with the PT cell. Thus PST studies can help elucidate anatomical connections.

Modeling the processes underlying the PST histogram is exceedingly complicated. To see this, let us examine what must be the simplest possible, and hence unrealistic, paradigm arrangement. One cell (literally) is to receive a stimulus at time $t = 0$. This cell connects monosynaptically with and excites another cell from which recordings of spikes will be made. Then the latency is

$$T_L = T_1 + T_2 + T_3 + T_4, \tag{10.163}$$

where T_1 is the time taken to elicit an action potential in the stimulated cell, T_2 is the conduction time of an action potential along an axon to its terminals, T_3 is the synaptic delay time, and T_4 is the time taken to elicit an action potential in the cell from which recordings are being made.

Of course, the situation is much more complicated if there are whole populations of cells being stimulated and these are connected through populations of interneurons, which may be excitatory or inhibitory, to the cell from which it is being recorded. To further complicate matters, if the target cell is spontaneously active, one has the additional problem of distinguishing the "signal" from the "noise." This process is broadly called *filtering* and an *adaptive* technique has been employed for PST histograms with seemingly excellent results by Sanderson (1980).

10.14 Statistical analysis of the simultaneous spiking activities of two or more neurons

All of the methods of analysis presented so far in this chapter have concerned the activity of a single cell. However, it is often of interest to examine the ongoing activities of several cells in a given experimental situation. Collections of neurons that might be influencing each other or are affected by common sources of stimulation are often referred to as *neural networks*. Understanding the functioning of neural networks is one of the principal aims of theoretical neurobiology.

If it is possible to record intracellularly from more than one cell at the same time, then statistical analysis may proceed directly. This was the case, for example, in studies of identified cells of the abdominal ganglion of *Aplysia californica* by Bryant, Marcos, and Segundo (1973).

Extracellular recordings of the spikes of several neurons may be made with one or several electrodes, with perhaps as many as 24 recording channels (Kuperstein and Whittington 1981). If there are many electrodes sufficiently separated in space, it may be possible to record the individual spike trains of different cells. As with intracellular recording, in this fortunate circumstance an investigation of the statistical relation between the spike trains may be undertaken. An example is provided by the work of Johnson and Kiang (1976) on pairs of auditory nerve fibers in cats.

10.14.1 The separation problem

However, a less fortunate situation will often exist; namely, the spikes from several cells are recorded in the one channel or from just one electrode. It is then a challenging statistical problem to determine which sets of spikes have emanated from particular cells. Before this is done, it may be advantageous or necessary to separate spikes from background noise. A method of so producing a "string of clean spikes" by utilizing the fact that the time derivatives of the potential during a spike are often quite large has been illustrated in experiments on moth receptor cells involved in pheromone detection (Rumbo 1980).

To separate superimposed spike trains into their components, suitable metrics are required to quantify the differences between spikes in given classes. One separation technique, based on latency measurements, has been employed to analyze the activity of cat somatosensory cells and hence to assist in the determination of their receptive fields (Looft and Fuller 1979). Another method of discrimination is to sort spikes according to their amplitudes (Kovbasa, Nozdrachev, and Yagodin 1985) but this is made hazardous by the variability in amplitudes of spikes recorded from a given cell. A finer discrimination may be attained by using entire spike waveforms. This was achieved automatically with a set of *templates* for spikes of cells involved with the shadow reflex in barnacles by Millecchia and McIntyre (1978). Some of their results are shown in Figure 10.4. Another study, also based on shapes, construed the separation as a pattern recognition problem and developed an algorithm for reducing the number of features needed in comparing spikes and achieving a discrimination (Dinning and Sanderson 1981; see this paper for several references to earlier work).

It is perhaps worth mentioning that a well-known distance function, the L_2 metric, can be used for two waveforms $f(t)$ and $g(t)$ defined on the same interval (a, b). This is given by

$$d(f, g) = \int_a^b (f(t) - g(t))^2 \, dt,$$

and may be approximated by a version at a discrete set of time points

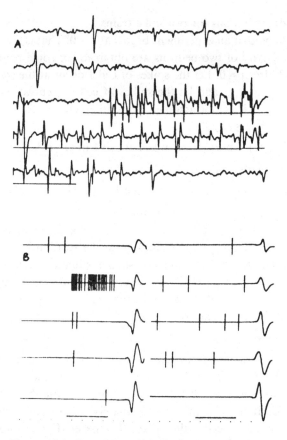

Figure 10.4. Separation of individual spike trains from a single extracellular recording. A – The original record with the stimulus (shadow) applies during the solid accompanying straight line. B – Showing 10 templates and corresponding spike trains extracted from the recording in A. Not all spikelike events are successfully extracted. [Adapted from Millecchia and McIntyre (1978).]

$t_1, \ldots, t_n$ as follows:

$$\tilde{d}(f, g) = \sum_{k=1}^{n} \left(f(t_k) - g(t_k) \right)^2.$$

Let us assume that sequences of spike times from one or more cells have been obtained. The temporal relationships between the spike trains may be obvious and a statistical investigation not warranted. For example, examination of the first two trains in Figure 10.4B shows that the first cell was made silent and the second emitted a high-frequency burst of spikes during the presentation of the stimulus. In many cases, however, things are not that simple and statistical testing is desirable.

10.14.2 Basic definitions for two spike trains

Assume that an experiment begins at $t = 0$. That is, records of the spiking activity of two neurons are obtained for some reasonable time interval after $t = 0$. Let the spikes of cell 1 occur at the (random) times $\Theta_{11}, \Theta_{12}, \Theta_{13}, \ldots$ and let the spikes of cell 2 occur at the times $\Theta_{21}, \Theta_{22}, \Theta_{23}, \ldots$. It is not assumed that either Θ_{11} or Θ_{21} are zero. Define also, for $t_1 \geq t_2 \geq 0$,

$$N_1(t_1, t_2) = \text{number of spikes of cell 1 in } (t_1, t_2], \quad (10.164\text{A})$$

with the abbreviated notation

$$N_1(t) = \text{number of spikes of cell 1 in } (0, t], \quad (10.164\text{B})$$

with $N_1(0) = 0$ and similarly for neuron 2.

Before proceeding with any statistical analysis, a *visual inspection* of the two spike trains may reveal clues as to the relation between the cells. For example, in an extremely simple situation, cell 1 might monosynaptically excite cell 2 and every spike in cell 1 might be followed after a delay of Δ (possibly random) by a spike in cell 2. Then the following relation would hold:

$$\Theta_{2,k} = \Theta_{1,k} + \Delta. \quad (10.165)$$

Thus scatter diagrams may be helpful. Apart from the usual plot of the points $(\Theta_{1k}, \Theta_{2k})$, $k = 1, 2, \ldots$, it may be useful to plot the ordinal number of the spike of neuron 1 that preceded the kth spike of neuron 2. For it may be that approximately a constant number of spikes of one cell precedes the spike of another. For example, the relation

$$\Theta_{2,k} = \Theta_{1,3k} + \Delta \quad (10.166)$$

would indicate that three spikes in cell 1 were followed after a delay by a spike in cell 2, suggesting the presence of a connection between the two cells.

The joint spike density

In complex networks, however, obvious and simple relations such as the above are less likely to be found and one needs a repertoire of statistical concepts and tests. The *joint spike density* (or product density), denoted by a $u(t_1, t_2)$, is defined as

$$u(t_1, t_2) = \lim_{\Delta t_1, \Delta t_2 \to 0^+} \frac{\Pr\{\text{a spike from neuron 1 occurs in } (t_1, t_1 + \Delta t_1] \text{ and a spike from neuron 2 occurs in } (t_2, t_2 + \Delta t_2]\}}{\Delta t_1 \Delta t_2}, \quad t_1, t_2 > 0.$$

$$(10.167)$$

Note that this involves an unconditional joint probability and that if the spike trains are independent

$$u(t_1, t_2) = u_1(t_1)u_2(t_2), \tag{10.168}$$

where u_1 and u_2 are the individual spike densities. For example, if both the spike trains are independent and Poisson with intensities λ_1 and λ_2, then $u(t_1, t_2) = \lambda_1\lambda_2$, constant for all t_1 and t_2. If the simple relation (10.165) exists, then $u(t_1, t_2)$ will be concentrated on the line $t_2 = t_1 + \Delta$.

We may *estimate* $u(t_1, t_2)$ as follows. Suppose that an experiment is performed m times in such a way that there is no interference between trials and each run has duration s. Let the spike times of neuron 1 on trial l be $\Theta_{1,k}^{(l)}$, $k = 1, \ldots, m_l$, and those of neuron 2 be $\Theta_{2,k}^{(l)}$, $k = 1, \ldots, n_l$, where in both cases $l = 1, \ldots, m$. This gives on trial l a total number of $m_l n_l$ data points in the square $S = [0, s] \times [0, s]$. By subdividing $[0, s]$ into a certain number of equal subintervals of length b, the square S is subdivided into squares of area b^2. The value of b is chosen so that on any trial not more than one of the $m_l n_l$ points falls in any given square. The *joint spike density may now be estimated by*

$$\hat{u}\left(\frac{ib}{2}, \frac{jb}{2}\right) = \frac{\begin{array}{c}\text{number of times in } m \text{ trials a data point fell}\\ \text{in a square centered on } \left(\dfrac{ib}{2}, \dfrac{jb}{2}\right)\end{array}}{mb^2}$$

$$\tag{10.169}$$

Hypotheses concerning $u(t_1, t_2)$ can be examined using a χ^2-goodness-of-fit test.

Cross-intensity functions

Assuming at least second-order stationarity, we define for spike trains 1 and 2, the *cross-intensity functions*:

$$u_{12}(t) = \lim_{\Delta t \to 0} \frac{1}{\Delta t} \Pr\{\text{neuron 2 spikes in}$$

$$\times (s + t, s + t + \Delta t] | \text{neuron 1 spikes at } s\}, \tag{10.170}$$

$$u_{21}(t) = \lim_{\Delta t \to 0} \frac{1}{\Delta t} \Pr\{\text{neuron 1 spikes in}$$

$$\times (s + t, s + t + \Delta t] | \text{neuron 2 spikes at } s\}. \tag{10.171}$$

Here $t \neq 0$ but t can be positive or negative.

Utilizing the definition of conditional probability and assuming the separate spike trains are stationary so that the individual spike densities are given by

$$u_k(t) = 1/\mu_k, \qquad k = 1, 2, \tag{10.172}$$

we find that

$$\frac{u_{12}(t)}{\mu_1} = \frac{u_{21}(-t)}{\mu_2}, \tag{10.173}$$

as noted by Perkel, Gerstein, and Moore (1967b). This means that u_{21} is known if u_{12} is known and vice versa, provided, of course, the mean intervals are known.

The cross-intensity function has often been called the *cross-correlation function* and its estimator a *cross-correlation histogram*. To estimate $u_{12}(t)$ assume the spike train 1 consists of the points $\Theta_{1,k}$, $k = 1, \ldots, m$, and spike train 2 consists of the points $\Theta_{2,k}$, $k = 1, \ldots, n$. A suitable bin width b is chosen. Choose $\Theta_{1,1}$ as origin and ascertain which bin $B_j = (jb, (j+1)b)$, $j = 0, \pm 1, \pm 2, \ldots$, each of the n spikes in the second train falls relative to this origin. Repeat this using each of the m spikes in the first-spike train as origin so that eventually mn points have been allocated. Let v_j be the total number of points in m "trials" that fell in B_j. Then an estimator for $u_{12}((j + \frac{1}{2})b)$ is

$$\hat{u}_{12}\left(\left(j + \frac{1}{2}\right)b\right) = \frac{v_j}{bm}. \tag{10.174}$$

Many such experimental histograms have been reported [see, for example, Perkel et al. 1967b, Moore et al. (1970), Bryant et al. (1973), Toyama, Kimura, and Tanaka (1981), and Toyama and Tanaka (1984)]. An example is given in Figure 10.5. See Brillinger (1986b) for further references.

The cross periodogram

One may define the *Fourier transform* of each spike train as a stochastic integral

$$J_j(\omega) = \int_0^\infty e^{i\omega t} \, dN_j(t), \qquad j = 1, 2,$$

$$= \sum_k e^{i\omega \Theta_{j,k}}, \tag{10.175}$$

where ω is frequency and summation is over all spikes. Thus one may define further the cross periodogram

$$I_{12}(\omega) = J_1(\omega)\overline{J_2(\omega)}, \tag{10.176}$$

where the overbar indicates complex conjugate. The use of this

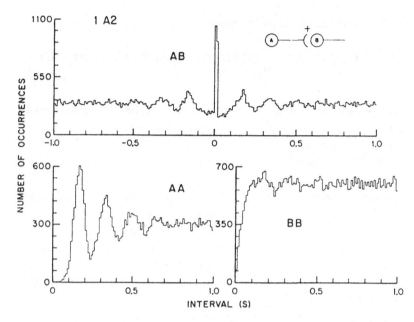

Figure 10.5. An example of a histogram estimating the cross-intensity function of the two spike trains from two neurons (labeled A and B) of *Aplysia californica*. The two lower histograms are the individual spike-density estimates. The inset shows the synaptic connection as monosynaptic with A exciting B. [Adapted from Moore et al. (1970).]

quantity to analyze bivariate point processes was advocated by Jenkins [in Bartlett (1963)] and such methods have been developed by Brillinger (1975, 1986a, b) who also utilizes the *coherence* and *cross spectrum*. We will not go into any details of these frequency-domain methods but refer the reader to the above sources as well as Brillinger, Bryant, and Segundo (1976).

10.14.3 Testing for independence

Given that the spike-time sequences of two cells do not reveal an obvious relationship and that one suspects a more subtle relationship may exist, a first step should consist of testing for the *independence* or otherwise of the two trains of impulses. This is a crucial step whose importance cannot be overemphasized. We will outline two approaches that have been taken; the first is based on cross-intensity functions and the second on recurrence times.

Examination of the cross-intensity function

Consider the cross-intensity function defined by (10.170). If the two spike trains are independent, then the conditioning event will be irrelevant and $v_{12}(t)$ will reduce to

$$u_{12}(t) = \lim_{\Delta t \to 0} \frac{1}{\Delta t} \Pr\{\text{neuron 2 spikes in } (s+t, s+t+\Delta t]\}.$$

(10.177)

This is just the quantity $u_2(t)$ that we have called the spike density for neuron 2. If the spike train of neuron 2 is, in fact, stationary or an equilibrium renewal process, then the cross-intensity function will be constant. This should be emphasized since on reading the literature the impression may be gained that independence alone implies a constant value for u_{12} or u_{21}. [Note that in the case of independence (10.173) must become the identity $(\mu_1\mu_2)^{-1} = (\mu_1\mu_2)^{-1}$ because stationarity was already assumed.] Thus, in using the cross-intensity function u_{12} to test for independence, one must compare this quantity with the spike intensity of the individual train u_2. Indeed, experimentalists often report all of u_{12}, u_2 and u_1 as in Moore et al. (1970) and Bryant et al. (1973). One such set of results is shown in Figure 10.5 for two cells of the visceral ganglion of *Aplysia*.

There are no critical values available to ascertain whether a departure of u_{12} from u_2 is significant. However, some special cases have been treated theoretically by Perkel et al. (1967b) and empirical approaches suggested by Lewis were reported by Bryant et al. (1973).

Recurrence times

We will define and summarize the relevant facts about *recurrence times*, explain briefly how they have been employed to test for independence, and raise a question as to the applicability of the test. Further theoretical details are given in Cox (1962), McFadden (1962), and Cox and Miller (1965). One of the first experimental studies by Wyman (1966) on the neural mechanisms of insect flight used a different terminology. Let t be an arbitrary but fixed time point.

The *forward recurrence time* V_t is defined as the waiting time to the next spike after t. Then, clearly, V_t can take values between 0 and ∞. The *backward recurrence time* U_t is defined as the time interval back from t to the most recent spike. If the experiment started at $t = 0$, then U_t must lie in the interval $[0, t]$. If a *stationary-point process* is being observed, then it is usual to assume it has been running since $t = -\infty$ and hence in that case both U_t and V_t take values in $(0, \infty)$. Of course, in the limit $t \to \infty$ the distinction as to the starting time is not necessary.

For the $t \to \infty$ case of a renewal process starting at 0 (Cox 1962) and always for a stationary-point process (McFadden 1962), the probability density function of the forward and backward recurrence times may be shown to be given by

$$f(x) = \frac{1 - F(x)}{\mu}, \qquad 0 < x < \infty, \tag{10.178}$$

where $F(x)$ is the distribution function of the time interval between spikes and where μ is the mean interval. Note that the interspike intervals are not required to be independent.

The test for independence of two spike trains has been implemented as follows (Moore et al. 1966; Perkel et al. 1967b; Johnson and Kiang 1976). Measure the time intervals between spikes in train 1 and the nearest future spike in train 2. This gives the *forward cross-interval* histogram, which collects values of the random variables

$$V_k = \min_j \left(\Theta_{2,j} - \Theta_{1,k} \right), \qquad \Theta_{2,j} > \Theta_{1,k}, \qquad k = 1, 2, \ldots . \tag{10.179}$$

Similarly, measure the time intervals from spikes in train 1 to the most recent of the prior spikes in train 2. This gives the *backward cross-interval* histogram, which collects values of the random variables

$$U_k = \min_j \left(\Theta_{1,k} - \Theta_{2,j} \right), \qquad \Theta_{2,j} \leq \Theta_{1,k}, \qquad k = 1, 2, \ldots . \tag{10.180}$$

It is supposed that the U_k's and the V_k's should have the distributions of the backward and forward recurrence times. Thus the histogram of values of the cross intervals may be compared with the density (10.178) by using the experimental interspike-interval distribution. If there is not a significant discrepancy, one suspects the spike trains are independent. However, the recurrence times are based on a fixed yet arbitrary reference time, *not* a random one. The use of a random time of occurrence train 1 as a reference time for train 2 is suggestive of a *strong Markov assumption* except that random times in train 1 cannot be *stopping times* for train 2, unless there is a purely deterministic relation between them [see, for example, Kannan (1979), for the definitions of these terms]. It is not clear that the fixed but arbitrary time used to define the recurrence times can be replaced by a randomly selected time.

Finally, we point to some further studies on the analysis of statistical relations between spike trains. The shape of the cross-intensity function has been used as a clue to the possible nature of the relation between neurons. A systematic investigation was performed by Perkel et al. (1967b) and Moore et al. (1970). A theoretical study was made by Knox (1974). The studies on cells in the visual cortex by Toyama and Tanaka (1984) have been most interesting and have revealed the mechanisms involved in the generation of the various kinds of receptive fields. Brillinger et al. (1976) and Brillinger (1986a, b) have given a method for analyzing the interrelationships between three neurons. Habib and Sen (1985) have discussed methods for the analysis of nonstationary spike trains. A methodology for input–output relations using Wiener kernel-type expansions for point processes has been developed by Brillinger (1975), Krausz (1975), Brillinger et al. (1976), and Krausz and Friesen (1977).

REFERENCES

Abramowitz, M., and Stegun, I. (eds.) (1965). *Handbook of Mathematical Functions*. Dover, New York.

Adelman, W. J., and Fitzhugh, R. (1975). Solutions of the Hodgkin–Huxley equations modified for potassium accumulation in a periaxonal space. *Fed. Proc.* **34** 1322–9.

Ames, W. F. (1977). *Numerical Methods for Partial Differential Equations*. Academic, New York.

Araki, T., and Terzuolo, C. A. (1962). Membrane currents in spinal motorneurons associated with the action potential and synaptic activity. *J. Neurophysiol.* **25** 772–89.

Aronson, D. G. (1978). A comparison method for stability analysis of nonlinear parabolic problems. *SIAM Rev.* **20** 245–64.

Aronson, D. G., and Weinberger, H. F. (1975). Nonlinear diffusion in population genetics, combustion and nerve propagation. In *Partial Differential Equations and Related Topics*. Springer, New York.

Ash, R. B. (1970). *Basic Probability Theory*. Wiley, New York.

Baker, M. A. (1971). Spontaneous and evoked activity of neurones in the somatosensory thalamus of the waking cat. *J. Physiol.* **217** 359–79.

Barlow, R. F., and Proschan, F. (1975). *Statistical Theory of Reliability and Life Testing*. Holt, Rinehart, and Winston, New York.

Barrett, J. N., and Crill, W. E. (1974). Influence of dendritic location and membrane properties on the effectiveness of synapses on cat motoneurones. *J. Physiol.* **239** 325–45.

Bartlett, M. S. (1963). The spectral analysis of point processes. *J. Roy. Statist. Soc. B.* **25** 264–96.

Bassant, M. H. (1976). Analyse statistique de l'activite des cellules pyramidales de l'hippocampe dorsal du lapin. *Electroenceph. Clin. Neurophysiol.* **40** 585–603.

Bell, J. (1981). Modeling parallel, unmyelinated axons: pulse trapping and ephaptic transmission. *SIAM J. Appl. Math.* **41** 168–80.

Bell, J., and Cook, L. P. (1978). On the solutions of a nerve conduction equation. *SIAM J. Appl. Math.* **35** 678–88.

Bell, J., and Cook, L. P. (1979). A model of the nerve action potential. *Math. Biosci.* **46** 11–36.

Bennett, M. R., and Florin, T. (1974). A statistical analysis of the release of acetylcholine at newly formed synapses in striated muscle. *J. Physiol.* **238** 93–107.

Benoit, O., and Chataignier, C. (1973). Patterns of spontaneous unitary discharge in thalamic ventrobasal complex during waking and sleep. *Exp. Brain Res.* **17** 348–63.

Billingsley, P. (1968). *Convergence of Probability Measures*. Wiley, New York.

Bishop, P. O., Levick, W. R., and Williams, W. O. (1964). Statistical analysis of the dark discharge of lateral geniculate neurones. *J. Physiol.* **170** 598–612.

Bornstein, J. C. (1978). Spontaneous multiquantal release at synapses in guinea-pig hypogastric ganglia: evidence that release can occur in bursts. *J. Physiol.* **282** 375–98.

Boyce, W. E., and DiPrima, R. C. (1977). *Elementary Differential Equations and Boundary Value Problems*. Wiley, New York.

Boyd, I. A., and Martin, A. R. (1956). Spontaneous subthreshold activity at mammalian neuromuscular junctions. *J. Physiol.* **132** 61–73.

Braitenberg, V. (1965). What can be learned from spike interval histograms about synaptic mechanisms? *J. Theor. Biol.* **8** 419–25.

Braitenberg, V., Gambardella, G., Ghigo, G., and Vota, U. (1965). Observations on spike sequences from spontaneously active Purkinje cells in the frog. *Kybernetik* **2** 197–205.

Breiman, L. (1968). *Probability*. Addison-Wesley, Reading, Mass.

Brill, M. H., Waxman, S. G., Moore, J. W., and Joyner, R. W. (1977). Conduction velocity and spike configuration in myelinated fibres: computed dependence on internode distance. *J. Neurol. Neurosurg. Psych.* **40** 769–74.

Brillinger, D. R. (1975). The identification of point process systems. *Ann. Probab.* **3** 909–29.

Brillinger, D. R. (1986a). Analysing interacting nerve cell spike trains to assess causal connections. *Proc. Conf. Advanced Methods of Physiological System Modeling*, USC.

Brillinger, D. R. (1986b). Some statistical methods for random process data from seismology and neurophysiology. Technical Report 84, Department of Statistics, University of California, Berkeley.

Brillinger, D. R., Bryant, H. L., and Segundo, J. P. (1976). Identification of synaptic interactions. *Biol. Cybernetics* **22** 213–28.

Brink, F., Bronk, D. W., and Larrabee, M. G. (1946). Chemical excitation of nerve. *Ann. N.Y. Acad. Sci.* **47** 457–85.

Bromm, B., Schwarz, J. R., and Ochs, G. (1981). A quantitative analysis of combined potential and current clamp experiments on the single myelinated nerve fibre of *Rana esculenta*. *J. Theoret. Neurobiol.* **1** 120–32.

Brown, M. C., and Stein, R. B. (1966). Quantitative studies on the slowly adapting stretch receptor of the crayfish. *Kybernetik* **3** 175–85.

Brown, T. H., Perkel, D., and Feldman, M. W. (1976). Evoked transmitter release: statistical effects of nonuniformity and nonstationarity. *Proc. Natl. Acad. Sci. U.S.A.* **73** 2913–17.

Bryant, H. L., Marcos, A. R., and Segundo, J. P. (1973). Correlations of neuronal spike discharges produced by monosynaptic connections and by common inputs. *J. Neurophysiol.* **36** 205–25.

Bryant, H. L., and Segundo, J. P. (1976). Spike initiation by transmembrane current: a white noise analysis. *J. Physiol.* **260** 279–314.

Buller, A. J., Nicholls, J. G., and Strom, G. (1953). Spontaneous fluctuations of excitability in the muscle spindle of the frog. *J. Physiol.* **122** 409–18.

Buno, W., Fuentes, J., and Segundo, J. P. (1978). Crayfish stretch-receptor organs: effects of length steps with and without perturbations. *Biol. Cybernetics* **31** 99–110.

Burke, R. E., and Rudomin, P. (1977). Spinal neurons and synapses. In *Handbook of Physiology, The Nervous System*, Volume 1, Section 1, Chapter 24 (E. R. Kandel, ed.). American Physiological Society, Bethesda.

Burke, W. (1957). Spontaneous potentials in slow muscle fibres of the frog. *J. Physiol.* **135** 511–21.

Burke, W., and Sefton, A. J. (1966a). Discharge patterns of principal cells and interneurones in lateral geniculate nucleus of rat. *J. Physiol.* **187** 201–12.

Burke, W., and Sefton, A. J. (1966b). Recovery of responsiveness of cells of lateral geniculate nucleus of rat. *J. Physiol.* **187** 213–29.

Burns, B. D., and Webb, A. C. (1976). The spontaneous activity of neurones in the cat's cerebral cortex. *Proc. Roy. Soc. Lond. B.* **194** 211–23.

Cairoli, R., and Walsh, J. B. (1975). Stochastic integrals in the plane. *Acta Math.* **134** 111–81.

Calvin, W. H., and Hartline, D. K. (1977). Retrograde invasion of lobster stretch receptor somata in control of firing rate and extra spike patterning. *J. Neurophysiol.* **40** 106–18.

Calvin, W. H., Howe, J. F., and Loeser, J. D. (1977). Ectopic repetitive firing in focally demyelinated axons and some implications for trigeminal neuralgia. In *Pain in the Trigeminal Region* (D. J. Anderson and B. Matthews, eds.). Elsevier, Amsterdam.

Calvin, W. H., and Stevens, C. F. (1965). A Markov process model for neuron behavior in the interspike interval. *Proc. 18th Annual Conf. Engineering in Medicine and Biology* **7** 118 (Abstract).

Calvin, W. H., and Stevens, C. F. (1968). Synaptic noise and other sources of randomness in motoneuron interspike intervals. *J. Neurophysiol.* **31** 574–87.

Calvin, W. H., and Sypert, G. (1976). Fast and slow pyramidal tract neurons: an intracellular analysis of their contrasting repetitive firing properties in the cat. *J. Neurophysiol.* **39** 420–33.

Capocelli, R. M., and Ricciardi, L. M. (1971). Diffusion approximation and first passage time problem for a model neuron. *Kybernetik* **8** 214–23.

Carpenter, G. A. (1977). A geometric approach to singular perturbation problems with application to nerve impulse equations. *J. Diff. Equations* **23** 335–67.

Carpenter, G. A. (1978). A mathematical analysis of excitable membrane phenomena. *Progress in Cybernetics and Systems Research* **3** 505–14.

Carpenter, G. A. (1981). Normal and abnormal signal patterns in nerve cells. In *Mathematical Psychology and Psychophysiology* (S. Grossberg, ed.). AMS/SIAM Symposium Series.

Casten, R. H., Cohen, H., and Lagerstrom, P. (1975). Perturbation analysis of an approximation to Hodgkin–Huxley theory. *Quart. Appl. Math.* **32** 365–402.

Chandrasekhar, S. (1943). Dynamical friction. II. The rate of escape of stars from clusters and the evidence for the operation of dynamical friction. *Astrophys. J.* **97** 263–73.

Chatfield, C. (1975). *The Analysis of Time Series: Theory and Practice.* Chapman and Hall, London.

Chen, K.-H. (1976). Existence and uniqueness theorems of the Hodgkin–Huxley model on the propagation of nerve impulses. *Chinese J. Math.* **4** 17–28.

Chung, K. L. (1979). *Elementary Probability Theory.* Springer, New York.

Clay, J. R., and Goel, N. S. (1973). Diffusion models for firing of a neuron with varying threshold. *J. Theor. Biol.* **39** 633–44.

Coddington, E., and Levinson, N. (1955). *Theory of Ordinary Differential Equations.* McGraw-Hill, New York.

Cohen, H. (1971). Nonlinear diffusion problems. In *Studies in Applied Mathematics,* Volume 7. The Mathematical Association of America.

Cohen, I., Kita, H., and Van der Kloot, W. (1973). Miniature end-plate potentials: evidence that the intervals are not fit by a Poisson distribution. *Brain Res.* **54** 318–23.

Cohen, I., Kita, H., and Van der Kloot, W. (1974a). The intervals between miniature end-plate potentials in the frog are unlikely to be independently or exponentially distributed. *J. Physiol.* **236** 327–39.

Cohen, I., Kita, H., and Van der Kloot, W. (1974b). The stochastic properties of spontaneous quantal release of transmitter at the frog neuromuscular junction. *J. Physiol.* **236** 341–61.

Cole, K. S. (1949). Dynamic electrical characteristics of the squid axon membrane. *Arch. Sci. Physiol.* **3** 253–8.

Colquhoun, D., and Hawkes, A. G. (1981). On the stochastic properties of single ion channels. *Proc. Roy. Soc. Lond.* B **211** 205–35.

Colquhoun, D., and Hawkes, A. G. (1982). On the stochastic properties of bursts of single ion channel openings and of clusters of bursts. *Philos. Trans. Roy. Soc. Lond.* B **300** 1–59.

Conway, E., Hoff, D., and Smoller, J. (1978). Large time behavior of solutions of nonlinear reaction–diffusion equations. *SIAM J. Appl. Math.* **35** 1–16.

Cooley, J., and Dodge, F. A. (1966). Digital computer solutions for excitation and propagation of the nerve impulse. *Biophys. J.* **6** 583–99.

Cooley, J., Dodge, F., and Cohen, H. (1965). Digital computer solutions for excitable membrane models. *J. Cell. Comp. Physiol.* **66** 99–109.

Cope, D. K., and Tuckwell, H. C. (1979). Firing rates of neurons with random excitation and inhibition. *J. Theor. Biol.* **80** 1–14.

Correia, M. J., and Landolt, J. P. (1977). A point process analysis of the spontaneous activity of anterior semicircular canal units in the anesthetized pigeon. *Biol. Cybernetics* **27** 199–213.

Cox, D. R. (1962). *Renewal Theory*. Methuen, London.

Cox, D. R., and Hinkley, D. V. (1975). *Theoretical Statistics*. Chapman and Hall, London.

Cox, D. R., and Lewis, P. A. W. (1966). *The Statistical Analysis of Series of Events*. Methuen, London.

Cox, D. R., and Miller, H. D. (1965). *The Theory of Stochastic Processes*. Wiley, New York.

Cox, D. R., and Oakes, D. (1983). *Analysis of Survival Data*. Chapman and Hall, London.

Crank, J., and Nicolson, P. (1947). A practical method for numerical evaluation of solutions for partial differential equations of the heat conduction type. *Proc. Camb. Philos. Soc.* **43** 50.

Cronin, J. (1977). Some mathematics of biological oscillations. *SIAM Rev.* **19** 100–38.

Curtis, D. R., and Eccles, J. C. (1960). Synaptic action during and after repetitive stimulation. *J. Physiol.* **150** 374–98.

Curtis, D. R., and Ryall, R. W. (1966). The synaptic excitation of Renshaw cells. *Exp. Brain Res.* **2** 81–96.

Darling, D. A. D., and Siegert, A. J. F. (1953). The first passage problem for a continuous Markov process. *Ann. Math. Statist.* **24** 624–39.

DeFelice, L. J. (1977). Fluctuation analysis in neurobiology. *Int. Rev. Neurobiol.* **20** 169–208.

del Castillo, J., and Katz, B. (1954). Quantal components of the endplate potential. *J. Physiol.* **124** 560–73.

Dinning, G. J., and Sanderson, A. C. (1981). Real-time classification of multiunit neural signals using reduced feature sets. *IEEE Trans. Biomed. Eng.* **BME-28** 804–8.

Dodge, F. A., and Cooley, J. (1973). Action potential of the motorneuron. *IBM J. Res. Devel.* **17** 219–29.

Dodge, F. A., and Frankenhaeuser, B. (1958). Membrane currents in isolated frog nerve fibre under voltage clamp conditions. *J. Physiol.* **143** 76–90.

Dodge, F. A., and Frankenhaeuser, B. (1959). Sodium currents in the myelinated nerve fibre of *Xenopus laevis* investigated with the voltage clamp technique. *J. Physiol.* **148** 188–200.

Dormont, J. F. (1972). Patterns of spontaneous unit activity in the ventrolateral thalamic nucleus of cats. *Brain Res.* **37** 223–39.

Dynkin, E. B. (1965). *Markov Processes*, Volume 1. Springer, Berlin.

Eccles, J. C. (1964). *The Physiology of Synapses*. Springer, New York.

References

Eccles, J. C. (1969). *The Inhibitory Pathways of the Central Nervous System*. Thomas, Springfield, Ill.

Eccles, J. C. (1977). *The Understanding of the Brain*. McGraw-Hill, New York.

Eccles, J. C., Eccles, R. M., Iggo, A., and Itô, M. (1961). Distribution of recurrent inhibition among motoneurons. *J. Physiol.* **159** 479–99.

Eidelberg, E., Goldstein, G. P., and Deza, L. (1967). Evidence for serotonin as a possible inhibitory transmitter in some limbic structures. *Exp. Brain Res.* **4** 73–80.

Eilbeck, J. C., Luzader, S. D., and Scott, A. C. (1981). Pulse evolution on coupled nerve fibres. *Bull. Math. Biol.* **43** 389–400.

Ekholm, A., and Hyvärinen, J. (1970). A pseudo-Markov model for series of neuronal spike events. *Biophys. J.* **10** 773–96.

Enright, J. T. (1967). The spontaneous neuron subject to tonic stimulation. *J. Theor. Biol.* **16** 54–77.

Evans, J. W. (1972a). Nerve axon equations. I. Linear approximations. *Indiana Univ. Math. J.* **21** 877–85.

Evans, J. W. (1972b). Nerve axon equations. II. Stability at rest. *Indiana Univ. Math. J.* **22** 75–90.

Evans, J. W. (1972c). Nerve axon equations. III. Stability of the nerve impulse. *Indiana Univ. Math. J.* **22** 577–93.

Evans, J. (1975). Nerve axon equations. IV. The stable and the unstable impulse. *Indiana Univ. Math. J.* **24** 1169–90.

Evans, J. W., and Feroe, J. (1977). Local stability theory of the nerve impulse. *Math. Biosci.* **37** 23–50.

Evans, J. W., and Shenk, N. (1970). Solutions to axon equations. *Biophys. J.* **10** 1090–1101.

Evarts, E. V. (1964). Temporal patterns of discharge of pyramidal tract neurons during sleep and waking in the monkey. *J. Neurophysiol.* **27** 152–171.

Fatt, P., and Katz, B. (1951). An analysis of the end-plate potential recorded with an intra-cellular electrode. *J. Physiol.* **115** 320–70.

Fatt, P., and Katz, B. (1952). Spontaneous subthreshold activity at motor nerve endings. *J. Physiol.* **117** 109–28.

Feigin, P. D. (1976). Maximum likelihood estimation for continuous-time stochastic processes. *Adv. Appl. Probab.* **8** 712–36.

Feller, W. (1966). *An Introduction to Probability Theory and Its Applications*, Volume 2. Wiley, New York.

Feller, W. (1968). *An Introduction to Probability Theory and Its Applications*, Volume 1. Wiley, New York.

Feroe, J. (1978). Temporal stability of solitary impulse solutions of a nerve equation. *Biophys. J.* **21** 103–10.

Firth, D. R. (1966). Interspike interval fluctuations in the crayfish stretch receptor. *Biophys. J.* **6** 201–15.

Fisher, R. A. (1937). The wave of advance of advantageous genes. *Ann. Eugen.* **7** 355–69.

Fitzhugh, R. (1960). Thresholds and plateaus in the Hodgkin–Huxley nerve equations. *J. Gen. Physiol.* **43** 867–96.

Fitzhugh, R. (1961). Impulses and physiological states in theoretical models of nerve membrane. *Biophys. J.* **1** 445–66.

Folks, J. L., and Chhikara, R. S. (1978). The inverse Gaussian distribution and its statistical application – a review. *J. Roy. Statist. Soc.* B **40** 263–89.

Frankenhaeuser, B. (1959). Steady state inactivation of sodium permeability in myelinated nerve fibres of *Xenopus laevis*. *J. Physiol.* **148** 671–6.

Frankenhaeuser, B. (1960). Sodium permeability in toad nerve and squid nerve. *J. Physiol.* **152** 159–66.

Frankenhaeuser, B. (1962a). Delayed currents in myelinated nerve fibres of *Xenopus laevis* investigated with voltage clamp technique. *J. Physiol.* **160** 40–5.

Frankenhaeuser, B. (1962b). Instantaneous potassium currents in myelinated nerve fibres of *Xenopus laevis. J. Physiol.* **160** 46–53.

Frankenhaeuser, B. (1962c). Potassium permeability in myelinated nerve fibres of *Xenopus laevis. J. Physiol.* **160** 54–61.

Frankenhaeuser, B. (1963a). A quantitative description of potassium current in myelinated nerve fibres of *Xenopus laevis. J. Physiol.* **169** 424–30.

Frankenhaeuser, B. (1963b). Inactivation of the sodium carrying mechanism in myelinated nerve fibres of *Xenopus laevis. J. Physiol.* **169** 445–51.

Frankenhaeuser, B., and Hodgkin, A. L. (1957). The action of calcium on the electrical properties of squid axons. *J. Physiol.* **137** 218–44.

Frankenhaeuser, B., and Huxley, A. F. (1964). The action potential of the myelinated nerve fibre of *Xenopus laevis* as computed on the basis of voltage clamp data. *J. Physiol.* **171** 302–15.

Frankenhaeuser, B., and Moore, L. E. (1963a). The effects of temperature on the sodium and potassium permeability changes in myelinated nerve fibres of *Xenopus laevis. J. Physiol.* **169** 431–7.

Frankenhaeuser, B., and Moore, L. E. (1963b). The specificity of the initial current in myelinated nerve fibres of *Xenopus laevis.* Voltage clamp experiments. *J. Physiol.* **169** 438–44.

Frankenhaeuser, B., and Waltman, B. (1959). Membrane resistance and conduction velocity of large myelinated nerve fibres from *Xenopus laevis. J. Physiol.* **148** 677–82.

Gebber, G. L. (1975). The probabilistic behavior of central 'vasomotor' neurons. *Brain Res.* **96** 142–6.

Geisler, C. D., and Goldberg, J. M. (1966). A stochastic model of the repetitive activity of neurons. *Biophys. J.* **6** 53–69.

Gerstein, G. L., Butler, R. A., and Erulkar, S. D. (1968). Excitation and inhibition in cochlear nucleus. I. Tone-burst stimulation. *J. Neurophysiol.* **31** 526–36.

Gerstein, G. L., and Kiang, N. Y.-S. (1960). An approach to the quantitative analysis of electrophysiological data from single neurons. *Biophys. J.* **1** 15–28.

Gerstein, G. L., and Mandelbrot, B. (1964). Random walk models for the spike activity of a single neuron. *Biophys. J.* **4** 41–68.

Gihman, I. I., and Skorohod, A. V. (1972). *Stochastic Differential Equations.* Springer, Berlin.

Gihman, I. I., and Skorohod, A. V. (1974). *The Theory of Stochastic Processes. I.* Springer, Berlin.

Gihman, I. I., and Skorohod, A. V. (1975). *The Theory of Stochastic Processes. II.* Springer, Berlin.

Gihman, I. I., and Skorohod, A. V. (1976). *The Theory of Stochastic Processes. III.* Springer, Berlin.

Gluss, B. (1967). A model for neuron firing with exponential decay of potential resulting in diffusion equations for probability density. *Bull. Math. Biophys.* **29** 233–43.

Goel, N. S., Richter-Dyn, N., and Clay, J. R. (1972). Discrete stochastic models for firing of a neuron. *J. Theor. Biol.* **34** 155–84.

Goel, N. S., and Richter-Dyn, N. (1974). *Stochastic Models in Biology.* Academic, New York.

Goldberg, J. M., Adrian, H. O., and Smith, F. D. (1964). Response of neurons of the superior olivary complex of the cat to acoustic stimuli of long duration. *J. Neurophysiol.* **27** 706–49.

Goldberg, J. M., Fernandez, C., and Smith, C. E. (1982). Responses of vestibular-nerve afferents in the squirrel monkey to externally applied galvanic currents. *Brain Res.* **252** 156–160.

Goldberg, J. M., Smith, C. E., and Fernandez, C. (1984). Relation between regularity and responses to externally applied galvanic currents in vestibular nerve afferents of the squirrel monkey. *J. Neurophysiol.* **51** 1236–56.

Goldstein, S. S., and Rall, W. (1974). Changes of action potential shape and velocity for changing core conductor geometry. *Biophys. J.* **14** 731–57.

Gradshteyn, I. S., and Ryzhik, I. M. (1965). *Table of Integrals, Series and Products.* Academic, New York.

Granit, R. (1963). Recurrent inhibition as a mechanism of control. In *Brain Mechanisms* (G. Moruzzi, A. Fessard, and H. H. Jasper, eds.), pages 23–37. Elsevier, Amsterdam.

Granit, R. (1970). *The Basis of Motor Control.* Academic, New York.

Grossman, R. G., and Viernstein, L. J. (1961). Discharge patterns of neurons in cochlear nucleus. *Science* **134** 99–101.

Guttman, R., and Barnhill, R. (1970). Oscillation and repetitive firing in squid axons. Comparison of experiments with computations. *J. Gen. Physiol.* **55** 104–18.

Guttman, R., Feldman, L., and Lecar, H. (1974). Squid axon membrane response to white noise stimulation. *Biophys. J.* **14** 941–55.

Habib, M., and Sen, P. K. (1985). Non-stationary stochastic point-process models in neurophysiology with applications to learning. In *Biostatistics: Statistics in Biomedical, Public Health and Environmental Sciences* (P. K. Sen, ed.). Elsevier, Amsterdam.

Hafner, D., Borchard, U., Richter, O., and Neugebauer, M. (1981). Parameter estimation in Hodgkin–Huxley-type equations for membrane action potentials in nerve and heart muscle. *J. Theor. Biol.* **91** 321–45.

Hagiwara, S. (1954). Analysis of interval fluctuation of the sensory nerve impulse. *Japan J. Physiol.* **4** 234–40.

Hagiwara, S., and Oomura, Y. (1958). The critical depolarization for the spike in the squid giant axon. *Japan. J. Physiol.* **8** 234.

Hagiwara, S., and Saito, N. (1959). Voltage–current relations in nerve cell membrane of Orchidium vernaculum. *J. Physiol.* **148** 161–79.

Hanson, F. B., and Tuckwell, H. C. (1983). Diffusion approximations for neuronal activity including synaptic reversal potentials. *J. Theoret. Neurobiol.* **2** 127–53.

Hassard, B. (1978). Bifurcations of periodic solutions of the Hodgkin–Huxley model for the squid giant axon. *J. Theor. Biol.* **71** 401–20.

Hastings, S. P. (1975). Some mathematical problems from neurobiology. *Amer. Math. Monthly* **82** 881–95.

Hastings, S. P. (1976a). On the existence of homoclinic and periodic orbits for the Fitzhugh–Nagumo equations. *Quart. J. Math. Oxford* **27** 123–34.

Hastings, S. P. (1976b). On travelling wave solutions of the Hodgkin–Huxley equations. *Arch. Rational Mech. Anal.* **60** 229–57.

Hille, B. (1984). *Ionic Channels of Excitable Membranes.* Sinauer, Sunderland, Mass.

Hodgkin, A. L. (1951). The ionic basis of electrical activity in nerve and muscle. *Biol. Rev.* **26** 339–409.

Hodgkin, A. L., and Huxley, A. F. (1952a). Currents carried by sodium and potassium ions through the membrane of the giant axon of Loligo. *J. Physiol.* **116** 449–72.

Hodgkin, A. L., and Huxley, A. F. (1952b). The components of membrane conductance in the giant axon of Loligo. *J. Physiol.* **116** 473–96.

Hodgkin, A. L., and Huxley, A. F. (1952c). The dual effect of membrane potential on sodium conductance in the giant axon of Loligo. *J. Physiol.* **116** 497–506.

Hodgkin, A. L., and Huxley, A. F. (1952d). A quantitative description of membrane current and its application to conduction and excitation in nerve. *J. Physiol.* **117** 500–44.

Hodgkin, A. L., Huxley, A. F., and Katz, B. (1952). Measurement of current–voltage relations in the membrane of the giant axon of *Loligo. J. Physiol.* **116** 424–48.

Hodgkin, A. L., and Katz, B. (1949). The effect of sodium ions on the electrical activity of the giant axon of the squid. *J. Physiol.* **108** 37–77.

Hodgson, J. P. E. (ed.) (1983). *Oscillations in Mathematical Biology.* Springer, Berlin.

Holden, A. V. (1976a). The response of excitable membrane models to a cyclic input. *Biol. Cybernetics* **21** 1–7.

Holden, A. V. (1976b). *Models of the Stochastic Activity of Neurones.* Springer, New York.

Holden, A. V. (1980). Autorhythmicity and entrainment in excitable membranes. *Biol. Cybernetics* **38** 1–8.

Holden, A. V. (1981). Membrane current fluctuations and neuronal information processing. In *Advances in Physiological Sciences 30: Neural Communications and Control* (G. Szekely ed.). Pergamon, Oxford.

Holden, A. V. (1982). The mathematics of excitation. In *Biomathematics in 1980* (L. M. Ricciardi and A. C. Scott, eds.). North-Holland, Amsterdam.

Holden, A. V. (1983). Stochastic processes in neurophysiology: transformation from point to continuous processes. *Bull. Math. Biol.* **45** 443–65.

Holden, A. V., and Ramadan, S. M. (1979). Stable distributions of interspike interval obtained from neurones of the pond snail. *J. Physiol.* **290** 26–27P.

Holden, A. V., and Yoda, M. (1981). The effect of ionic channel density on neuronal function. *J. Theoret. Neurobiol.* **1** 60–81.

Hubbard, J. L., and Jones, S. F. (1973). Spontaneous quantal transmitter release: a statistical analysis and some implications. *J. Physiol.* **232** 1–21.

Hunter, P. J., McNaughton, P. A., and Noble, D. (1975). Analytical models of propagation in excitable cells. *Prog. Biophys. Molec. Biol.* **30** 99–144.

Huxley, A. F., and Stampfli, R. (1949). Evidence for saltatory conduction in peripheral myelinated nerve fibres. *J. Physiol.* **108** 315–39.

Itô, K. (1951). On stochastic differential equations. *Mem. Amer. Math. Soc.* **4**.

Jack, J. J. B., Noble, D., and Tsien, R. W. (1985). *Electrical Current Flow in Excitable Cells.* Clarendon, Oxford.

Jackson, M. B. (1985). Stochastic behavior of a many-channel membrane system. *Biophys. J.* **47** 129–137.

Jansen, J. K. S., Nicolaysen, K., and Rudjord, T. (1966). Discharge patterns of neurons of the dorsalspinocerebellar tract activated by static extension of primary endings of muscle spindles. *J. Neurophysiol.* **29** 1061–86.

Jaswinski, A. H. (1970). *Stochastic Processes and Filtering Theory.* Academic, New York.

Johannesma, P. I. M. (1968). Diffusion models for the stochastic activity of neurons. In *Neural Networks* (E. R. Caianiello, ed.). Springer, Berlin.

Johnson, D. H., and Kiang, N. Y.-S. (1976). Analysis of discharges recorded simultaneously from pairs of auditory nerve fibers. *Biophys. J.* **16** 719–34.

Jones, C. K. R. T. (1984). Stability of the travelling wave solution of the Fitzhugh–Nagumo system. *Trans. Amer. Math. Soc.* **286** 431–69.

Junge, D., and Moore, G. P. (1966). Interspike-interval fluctuations in *Aplysia* pacemaker neurons. *Biophys. J.* **6** 411–34.

Kallianpur, G. (1983). On the diffusion approximation to a discontinuous model for a single neuron. In *Contributions to Statistics* (P. K. Sen, ed.). North-Holland, Amsterdam.

Kallianpur, G., and Wolpert, R. (1984a). Infinite dimensional stochastic differential equation models for spatially distributed neurons. *Appl. Math. Optim.* **12** 125–72.

Kallianpur, G., and Wolpert, R. (1984b). Weak convergence of solutions of stochastic differential equations with applications to nonlinear neuronal models. Technical Report 60, Department of Statistics, University of North Carolina, Chapel Hill.

Kannan, D. (1979). *An Introduction to Stochastic Processes.* North-Holland, Amsterdam.

Karlin, S., and Taylor, H. M. (1975). *A First Course in Stochastic Processes.* Academic, New York.

Katz, B., and Miledi, R. (1963). A study of miniature potentials in spinal motoneurones. *J. Physiol.* **168** 389–422.

Katz, B., and Schmidt, O. H. (1940). Electrical interaction between two nerve fibres. *J. Physiol.* **97** 471–88.

Keilson, J., and Ross, H. F. (1975). Passage time distributions for Gaussian Markov (Ornstein–Uhlenbeck) statistical processes. *Selected Tables in Mathematical Statistics* **3** 233–327.

Kernell, D. (1965). The adaptation and the relation between discharge frequency and current strength of cat lumbrosacral motoneurons stimulated by long-lasting injected currents. *Acta Physiol. Scand.* **65** 65–73.

Kernell, D., and Sjöholm, H. (1972). Motoneurone models based on "voltage clamp equations" for peripheral nerve. *Acta Physiol. Scand.* **87** 40–56.

Kernell, D., and Sjöholm, H. (1973). Repetitive impulse firing: comparisons between neurone models based on "voltage clamp equations" and spinal motoneurones. *Acta Physiol. Scand.* **87** 40–56.

Khodorov, B. I. (1974). *The Problem of Excitability.* Plenum, New York.

Khodorov, B. I., and Timin, E. N. (1975). Nerve impulse propagation along nonuniform fibres. *Prog. Biophys. Molec. Biol.* **30** 145–84.

Khodorov, B. I., Timin, E. N., Vilenkin, S. Ya., and Gul'ko, F. B. (1969). Theoretical analysis of the mechanisms of conduction of a nerve impulse over an inhomogeneous axon. I. Conduction through a portion with increased diameter. *Biophysics* **14** 323–35.

Kiang, N. Y.-S., and Moxon, E. C. (1974). Tails of tuning curves of auditory-nerve fibers. *J. Acoust. Soc. Am.* **55** 620–30.

Knox, C. K. (1974). Cross-correlation functions for a neuronal model. *Biophys. J.* **14** 567–82.

Koch, C., and Poggio, T. (1985). A simple algorithm for solving the cable equation in dendritic trees of arbitrary geometry. *J. Neurosci. Meth.* **12** 303–315.

Koch, C., Poggio, T., and Torre, V. (1983). Nonlinear interactions in a dendritic tree: localization, timing and role in information processing. *Proc. Natl. Acad. Sci. U.S.A.* **80** 2799–802.

Koch, C., Poggio, T., and Torre, V. (1986). Computations in the vertebrate retina: gain, enhancement, differentiation and motion discrimination. *Trends in Neurosci.* **9** 204–11.

Koike, H., Mano, N., Okada, Y., and Oshima, T. (1970). Repetitive impulses generated in fast and slow pyramidal tract cells by intracellularly applied current steps. *Exp. Brain Res.* **11** 263–81.

Kovbasa, S. I., Nozdrachev, A. D., and Yagodin, S. V. (1985). Statistical analysis of the multicellular activity of neurons. *J. Theoret. Neurobiol.* **4** 129–141.

Krausz, H. (1975). Identification of nonlinear systems using random impulse train inputs. *Biol. Cybernetics* **19** 217–30.

Krausz, H. I., and Friesen, W. O. (1977). The analysis of nonlinear synaptic transmission. *J. Gen. Physiol.* **70** 243–65.

Kryukov, V. I. (1976). Wald's identity and random walk models for neuron firing. *Adv. Appl. Probab.* **8** 257–77.

Kuffler, S. W., and Nicholls, J. G. (1976). *From Neuron to Brain.* Sinauer, Sunderland, Mass.

Kuperstein, M., and Whittington, D. A. (1981). ı\ practical 24 channel microelectrode for neural recording *in vivo. IEEE Trans. Biomed. Eng.* **BME-28** 288–93.

Kurtz, T. G. (1981). Approximation of discontinuous processes by continuous processes. In *Stochastic nonlinear Systems* (L. Arnold and R. Lefever, eds.). Springer, Berlin.

Kwaadsteniet, J. W. De. (1982). Statistical analysis and stochastic modelling of neuronal spike-train activity. *Math. Biosci.* **60** 17–71.

Lamarre, Y., Filion, M., and Cordeau, J. P. (1971). Neuronal discharges of the ventrolateral nucleus of the thalamus during sleep and wakefulness in the cat. I. Spontaneous activity. *Exp. Brain Res.* **12** 480–98.

Lampard, D. G., and Redman, S. J. (1969). Stochastic stimulation for the pharmacological study of monosynaptic spinal reflexes. *Eur. J. Pharmacol.* **5** 141–52.

Lansky, P. (1983). Inference for the diffusion models of neuronal activity. *Math. Biosci.* **67** 247–60.

Lansky, P. (1984). On approximations of Stein's neuronal model. *J. Theor. Biol.* **107** 631–47.

Lawless, J. F. (1982). *Statistical Models and Methods for Lifetime Data*. Wiley, New York.

Lebovitz, R. M. (1970). A theoretical examination of ionic interactions between neural and non-neural membranes. *Biophys. J.* **10** 423–44.

Lee, P. A. (1979). Some stochastic problems in neurophysiology. *S.E. Asian Bull. Math.* **11** 205–44.

Lees, M. (1967). An extrapolated Crank–Nicolson difference scheme for quasilinear parabolic equations. In *Nonlinear Partial Differential Equations* (W. F. Ames, ed.). Academic, New York.

Levick, W. R., Bishop, P. O., Williams, W. O., and Lampard, D. G. (1961). Probability distribution analyser programmed for neurophysiological research. *Nature* **192** 629–30.

Liptser, R. S., and Shiryayev, A. N. (1977). *Statistics of Random Processes I. General Theory*. Springer, New York.

Looft, F. J., and Fuller, M. S. (1979). Multiple unit correlation analysis of cutaneous receptors. *IEEE Trans. Biomed. Eng.* **BME-26** 572–8.

Losev, I. S. (1975). Model of the impulse activity of a neurone receiving a steady impulse influx. *Biofizika* **20** 893–900.

Losev, I. S., Shik, M. L., and Yagodnitsyn, A. S. (1975). Method of evaluating the synaptic influx to the single neurone of the mid-brain. *Biofizika* **20** 901–8.

Lovitt, W. V. (1950). *Linear Integral Equations*. Dover, New York.

MacGregor, R. J. (1968). A model for responses to activation by axodendritic synapses. *Biophys. J.* **8** 305–18.

MacGregor, R. M., and Lewis, E. R. (1977). *Neural Modeling*. Plenum, New York.

Marmarelis, P. Z., and Marmarelis, V. Z. (1978). *Analysis of Physiological Systems: The White Noise Approach*. Plenum, New York.

Marmont, G. (1949). Studies on the axon membrane, 1. A new method. *J. Cell. Comp. Physiol.* **34** 351–82.

Marsden, J. E., and McCracken, M. (1976). *The Hopf Bifurcation and Its Applications*. Springer, New York.

Martin, A. R. (1977). Junctional Transmission. II. Presynaptic mechanisms. In *Handbook of Physiology, The Nervous System*, Volume 1 (E. R. Kandel, ed.). American Physiological Society, Bethesda.

Matsuyama, Y., Shirai, K., and Akizuki, K. (1974). On some properties of stochastic information processes in neurons and neuron populations. *Kybernetik* **15** 127–45.

McFadden, J. A. (1962). On the lengths of intervals in a stationary point process. *J. Roy. Statist. Soc. B.* **24** 364–82.

McKean, H. P. (1970). Nagumo's equation. *Adv. Math.* **4** 209–23.

Millecchia, R., and McIntyre, T. (1978). Automatic nerve impulse identification and separation. *Comp. Biomed. Res.* **11** 459–68.

Miyamoto, M. D. (1975). Binomial analysis of quantal release at glycerol treated frog neuromuscular junctions. *J. Physiol.* **250** 121–42.

Molnar, C. E., and Pfeiffer, R. R. (1968). Interpretation of spontaneous spike discharge patterns of neurons in the cochlear nucleus. *Proc. IEEE* **56** 993–1004.

Moore, G. P., Perkel, D. H., and Segundo, J. P. (1966). Statistical analysis and functional interpretation of neuronal spike data. *Ann. Rev. Physiol.* **28** 493–522.

Moore, G. P., Segundo, J. P., Perkel, D. H., and Levitan, H. (1970). Statistical signs of synaptic interaction in neurons. *Biophys. J.* **10** 876–900.

Moore, J. W., Joyner, R. W., Brill, M. H., Waxman, S. D., and Najar-Joa, M. (1978). Simulations of conduction in uniform myelinated fibers, *Biophys. J.* **21** 147–60.

Morjanoff, M. P. (1971). Simulation studies on the impulse patterns in the discharge of single neurons and of small groups of interconnected neurons. MS Thesis, Monash University.

Murray, J. D. (1977). *Lectures on Nonlinear-Differential-Equation Models in Biology.* Clarendon, Oxford.

Muschaweck, L. G., and Loevner, D. (1978). Analysis of neuronal spike trains: survey of stochastic techniques and implementation of the cumulative-sums statistic for evaluation of spontaneous and driven activity. *Intern. J. Neurosci.* **8** 51–60.

Nagumo, J. S., Arimoto, S., and Yoshizawa, S. (1962). An active pulse transmission line simulating nerve axon. *Proc. I.R.E.* **50** 2061–70.

Nakahama, H., Nishioka, S., Otsuka, T., and Aikawa, S. (1966). Statistical dependency between interspike intervals of spontaneous activity in thalamic lemniscal neurons. *J. Neurophysiol.* **29** 921–34.

Nakahama, H., Suzuki, H., Yamamoto, M., Aikawa, S., and Nishioka, S. (1968). A statistical analysis of spontaneous activity of central single neurons. *Physiol. Behav.* **3** 745–52.

Nakamura, Y., Nakajima, S., and Grundfest, H. (1965). The action of tetrodotoxin on electrogenic components of squid giant axons. *J. Gen. Physiol.* **48** 985–96.

Neher, E. (1983). The charge carried by single channel currents of rat cultured muscle cells in the presence of local anaesthetics. *J. Physiol.* **339** 663–78.

Neher, E., and Stevens, C. F. (1977). Conductance fluctuations and ionic pores in membranes. *Ann. Rev. Biophys. Bioeng.* **6** 345–81.

Nilsson, H. G. (1977). Estimation of parameters in a diffusion neuron model. *Comp. Biomed. Res.* **10** 191–7.

O'Brien, J. H., Packham, S. C., and Brunnhoelzl, W. W. (1973). Features of spike train related to learning. *J. Neurophysiol.* **36** 1051–61.

Parnas, I., Hochstein, S., and Parnas, H. (1976). Theoretical analysis of parameters leading to frequency modulation along an inhomogeneous axon. *J. Neurophysiol.* **39** 909–23.

Parnas, I., and Segev, I. (1979). A mathematical model for conduction of action potentials along bifurcating axons. *J. Physiol.* **295** 323–43.

Parzen, E. (1962). *Stochastic Processes.* Holden-Day, San Francisco.

Pearson, E. S., and Hartley, H. O. (1972). *Biometrika Tables for Statisticians*, Volume 2. Cambridge University Press.

Pellet, J., Tardy, M.-F., Harlay, F., Dubrocard, S., and Gilhodes, J.-C. (1974). Activite spontanee des cellules de Purkinje chez le chat chronique: etude statistique des spikes complexes. *Brain Res.* **81** 75–95.

Perkel, D. H., Gerstein, G. L., and Moore, G. P. (1967a). Neuronal spike trains and stochastic point processes. *Biophys. J.* **7** 391–418.

Perkel, D. H., Gerstein, G. L., and Moore, G. P. (1967b). Neuronal spike trains and stochastic point processes. II. Simultaneous spike trains. *Biophys. J.* **7** 419–40.

Pfeiffer, R. R., and Kiang, N. Y.-S. (1965). Spike discharge patterns of spontaneous and continuously stimulated activity in the cochlear nucleus of anesthetized cats. *Biophys. J.* 5 301–16.

Poggio, G. F., and B. Viernstein, L. J. (1964). Time series analysis of impulse sequences of thalamic somatic sensory neurons. *J. Neurophysiol.* 27 517–45.

Poggio, T., and Torre, V. (1981). A theory of synaptic interactions. In *Theoretical Approaches in Neurobiology* (W. E. Reichhardt and T. Poggio, eds.). MIT Press, Cambridge, Mass.

Pollak, M., and Siegmund, D. (1985). A diffusion process and its applications to detecting a change in the drift of Brownian motion. *Biometrika* 72 267–80.

Pompeiano, O., and Wand, P. (1976). The relative sensitivity of Renshaw cells to static and dynamic changes of muscle length. In *Understanding the Stretch Reflex* (S. Homma, ed.), pages 199–222. Elsevier, Amsterdam.

Pyatigorskii, B. Ya. (1966). Steady nature of the spontaneous activity of the dorsal spino-cerebellum tract. *Biofizika* 11 116–22.

Ramon, F., Joyner, R. W., and Moore, J. W. (1975). Propagation of action potentials in inhomogeneous axon regions. *Fed. Proc.* 34 1357–63.

Redman, S. J., and Lampard, D. G. (1967). Monosynaptic stochastic stimulation of spinal motoneurones in the cat. *Nature* 216 921–2.

Redman, S. J., and Lampard, D. G. (1968). Monosynaptic stochastic stimulation of cat spinal motoneurons. I. Response of motoneurons to sustained stimulation. *J. Neurophysiol.* 31 485–98.

Redman, S. J., Lampard, D. G., and Annal, P. (1968). Monosynaptic stochastic stimulation of cat spinal motoneurons. II. Frequency transfer characteristics of tonically discharging motoneurons. *J. Neurophysiol.* 31 499–508.

Ricciardi, L. M. (1977). *Diffusion Processes and Related Topics in Biology*. Springer, Berlin.

Ricciardi, L. M., and Sacerdote, L. (1979). The Ornstein–Uhlenbeck process as a model for neuronal activity. *Biol. Cybernetics* 35 1–9.

Rinzel, J. (1975). Spatial stability of traveling wave solutions of a nerve conduction equation. *Biophys. J.* 15 975–88.

Rinzel, J. (1978). On repetitive activity in nerve. *Fed. Proc.* 37 2793–802.

Rinzel, J. (1979). Integration and propagation of neuroelectric signals. In *Studies in Mathematical Biology* 15 (S. A. Levin, ed.). The Mathematical Association of America.

Rinzel, J., and Keener, J. P. (1983). Hopf bifurcation to repetitive activity in nerve. *SIAM J. Appl. Math.* 43 907–22.

Rinzel, J., and Keller, J. B. (1973). Traveling wave solutions of a nerve conduction equation. *Biophys. J.* 13 1313–37.

Rinzel, J., and Miller, R. N. (1980). Numerical calculation of stable and unstable periodic solutions to the Hodgkin–Huxley equations. *Math. Biosci.* 49 27–59.

Rodieck, R. W., Kiang, N. Y.-S., and Gerstein, G. L. (1962). Some quantitative methods for the study of spontaneous activity of single neurons. *Biophys. J.* 2 351–68.

Roy, B. K., and Smith, D. R. (1969). Analysis of the exponential decay model of the neuron showing frequency threshold effects. *Bull. Math. Biophys.* 31 341–57.

Rumbo, E. R. (1980). Separation of interleaved nerve spike trains in a noisy channel. *IEEE Trans. Biomed. Eng.* BME-27 470–2.

Rushton, W. A. H. (1951). A theory of the effects of fibre size in medullated nerve. *J. Physiol.* 115 101–22.

Ryan, A., and Miller, J. (1977). Effects of behavioral performance on single unit firing patterns in inferior colliculus of the rhesus monkey. *J. Neurophysiol.* 40 943–56.

Sabah, N. H., and Leibovic, K. N. (1969). Subthreshold oscillatory responses of the Hodgkin–Huxley cable model for the squid giant axon. *Biophys. J.* 9 1206–22.

Sabah, N. H., and Spangler, R. A. (1970). Repetitive response of the Hodgkin–Huxley model for the squid giant axon. *J. Theor. Biol.* **29** 155–71.

Sampath, G., and Srinivasan, S. K. (1977). *Stochastic Models for Spike Trains of Single Neurons.* Springer, Berlin.

Sanderson, A. C. (1980). Adaptive filtering of neuronal spike train data. *IEEE Trans. Biomed. Engn.* **BME-27** 271–4.

Sato, S. (1978). On the moments of the firing interval of the diffusion approximated neuron. *Math. Biosci.* **39** 53–70.

Schmidt, D. S. (1976). Hopf's bifurcation theorem and the center theorem of Liapunov. In *The Hopf Bifurcation and Its Applications* (J. E. Marsden and M. McCracken, eds.). Springer, New York.

Schreiner, R. C., Essick, G. K., and Whitsel, B. L. (1978). Variability in somatosensory cortical neuron discharge: effects on capacity to signal different stimulus conditions using a mean rate code. *J. Neurophysiol.* **41** 338–49.

Scott, A. C. (1975). The electrophysics of a nerve fiber. *Rev. Mod. Phys.* **47** 487–533.

Scott, A. C. (1977). *Neurophysics.* Wiley, New York.

Scott, A. C., and Luzader, S. D. (1979). Coupled solitary waves in neurophysics. *Phys. Scripta* **20** 495–501.

Scott, A. C., and Vota-Pinardi, U. (1982). Pulse code transformations on axonal trees. *J. Theoret. Neurobiol.* **1** 173–95.

Segel, L. A. (ed.) (1980). *Mathematical Models in Molecular and Cellular Biology.* Cambridge University Press.

Siebert, W. M. (1965). Some implications of the stochastic behavior of primary auditory neurons. *Kybernetik* **2** 206–15.

Siegert, A. J. F. (1951). On the first passage time probability problem. *Phys. Rev.* **81** 617–23.

Silk, N., and Stein, R. B. (1966). Variability in the discharge frequency of single chemoreceptor fibres from the cat carotid body. *J. Physiol.* **186** 40P.

Shepherd, G. M., Brayton, R. K., Miller, J. P., Segev, I., Rinzel, J., and Rall, W. (1985). Signal enhancement in distal cortical dendrites by means of interactions between active dendritic spines. *Proc. Natl. Acad. Sci. U.S.A.* **82** 2192–5.

Skaugen, E. (1978). The effects of a finite number of sodium and potassium conducting pores upon firing behaviour in nerve models. *Inst. Informatikk*, University of Oslo, Oslo.

Smith, C. E., and Smith, M. V. (1984). Moments of voltage trajectories for Stein's model with synaptic reversal potentials. *J. Theoret. Neurobiol.* **3** 67–77.

Smith, D. R., and Smith, G. K. (1965). A statistical analysis of the continual activity of single cortical neurones in the cat unanasthetized isolated forebrain. *Biophys. J.* **5** 47–74.

Snyder, D. L. (1975). *Random Point Processes.* Wiley, New York.

Sparks, D. L., and Travis, R. P. (1968). Patterns of reticular unit activity observed during the performance of a discriminative task. *Physiol. Behav.* **3** 961–7.

Stein, R. B. (1965). A theoretical analysis of neuronal variability. *Biophys. J.* **5** 173–94.

Stein, R. B. (1967a). Some models of neuronal variability. *Biophys. J.* **7** 37–68.

Stein, R. B. (1967b). The frequency of nerve action potentials generated by applied currents. *Proc. Roy. Soc. Lond. B* **167** 64–86.

Stein, R. B., and Matthews, P. B. C. (1965). Differences in variability of discharge frequency between primary and secondary muscle spindle afferent endings of the cat. *Nature* **208** 1217–18.

Steriade, M., Wyzinski, P., and Apostol, V. (1973). Differential synaptic reactivity of simple and complex pyramidal tract neurons at various levels of vigilance. *Exp. Brain Res.* **17** 87–110.

Stevens, C. F. (1972). Inferences about membrane properties from electrical noise measurements. *Biophys. J.* **12** 1028–47.

Stevens, C. F. (1977). Study of membrane permeability changes by fluctuation analysis. *Nature* **270** 391–6.

Stratonovich, R. L. (1966). A new representation for stochastic integrals and equations. *J. SIAM Control.* **4** 362–71.

Stumpers, F. L. H. M. (1950). On a first-passage-time problem. *Philips Res. Rep.* **5** 270–81.

Sugiyama, H., Moore, G. P., and Perkel, D. H. (1970). Solutions for a stochastic model of neuronal spike production. *Math. Biosci.* **8** 323–41.

Swadlow, H. A., Kocsis, J. D., and Waxman, S. G. (1980). Modulation of impulse conduction along the axonal tree. *Ann. Rev. Biophys. Bioeng.* **9** 143–79.

Syka, J., Popelar, J., and Radil-Weiss, T. (1977). Comparison of spontaneous activity of mesencephalic reticular neurones in the waking state and during pentobarbital anaesthesia. *Physiol. Bohem.* **26** 21–30.

Tasaki, I., and Hagiwara, S. (1957). Demonstration of two stable potential states in the squid giant axon under tetraethylammonium chloride. *J. Gen. Physiol.* **40** 859–85.

Ten Hoopen, M. (1967). Pooling of impulse sequences, with emphasis on applications to neuronal spike data. *Kybernetik* **4** 1–10.

Thomas, M. U. (1975). Some mean first passage time approximations for the Ornstein–Uhlenbeck process. *J. Appl. Probab.* **12** 600–4.

Toyama, K., Kimura, M., and Tanaka, K. (1981). Cross-correlation analysis of inter-neuronal connectivity in cat visual cortex. *J. Neurophysiol.* **46** 191–201.

Toyama, K., and Tanaka, K. (1984). Visual cortical functions studied by cross-correla-tion analysis. In *Dynamic Aspects of Neocortical Function* (G. M. Edelman, W. E. Gall, and W. M. Cowan, eds.). Wiley, New York.

Traub, R. D. (1977a). Motoneurons of different geometry and the size principle. *Biol. Cybernetics* **25** 163–175.

Traub, R. D. (1977b). Repetitive firing of Renshaw spinal interneurons. *Biol. Cybernet-ics* **27** 71–6.

Traub, R. D. (1979). Neocortical pyramidal cells: a model with dendritic calcium conductance reproduces repetitive firing and epileptic behaviour. *Brain Res.* **173** 243–57.

Traub, R. D., and Llinás, R. (1977). The spatial distribution of ionic conductances in normal and axotomized motorneurons. *Neuroscience* **2** 829–49.

Traub, R. D., and Llinás, R. (1979). Hippocampal pyramidal cells: significance of dendritic ionic conductances for neuronal function and epileptogenesis. *J. Neuro-physiol.* **42** 476–96.

Troy, W. C. (1976). Bifurcation phenomena in Fitzhugh's nerve conduction equations. *J. Math. Anal. Appl.* **54** 678–90.

Troy, W. C. (1978). The bifurcation of periodic solutions in the Hodgkin–Huxley equations. *Quart. Appl. Math.* **36** 73–83.

Tsurui, A., and Osaki, S. (1976). On a first-passage problem for a cumulative process with exponential decay. *Stoch. Proc. Appl.* **4** 79–88.

Tuckwell, H. C. (1975). Determination of the interspike times of neurons receiving randomly arriving post-synaptic potentials. *Biol. Cybernetics* **18** 225–37.

Tuckwell, H. C. (1976a). On the first-exit time problem for temporally homogeneous Markov processes. *J. Appl. Probab.* **13** 39–48.

Tuckwell, H. C. (1976b). Firing rates of motoneurons with strong random synaptic excitation. *Biol. Cybernetics* **24** 147–52.

Tuckwell, H. C. (1976c). Frequency of firing of Stein's model neuron with application to cells of the dorsal spinocerebellar tract. *Brain Res.* **116** 323–8.

Tuckwell, H. C. (1977). On stochastic models of the activity of single neurons. *J. Theor. Biol.* **65** 783–5.

Tuckwell, H. C. (1978a). Neuronal interspike time histograms for a random input model. *Biophys. J.* **21** 289–90.

Tuckwell, H. C. (1978b). Recurrent inhibition and afterhyperpolarization: effects on neuronal discharge. *Biol. Cybernetics* **30** 115–23.

Tuckwell, H. C. (1979a). Synaptic transmission in a model for stochastic neural activity. *J. Theor. Biol.* **77** 65–81.

Tuckwell, H. C. (1979b). Solitons in a reaction–diffusion system. *Science* **205** 493–5.

Tuckwell, H. C. (1980a). Predictions and properties of a model of potassium and calcium ion movements during spreading cortical depression. *Int. J. Neurosci.* **10** 145–64.

Tuckwell, H. C. (1980b). Evidence for the existence of solitons in a nonlinear reaction–diffusion system. *SIAM J. Appl. Math.* **39** 310–22.

Tuckwell, H. C. (1981a). Simplified reaction–diffusion equations for spreading cortical depression. *Int. J. Neurosci.* **12** 95–107.

Tuckwell, H. C. (1981b). Poisson processes in biology. In *Stochastic Nonlinear Systems* (L. Arnold and R. Lefever, eds.). Springer, Berlin.

Tuckwell, H. C. (1984). Neuronal response to stochastic stimulation. *IEEE Trans. SMC* **14** 464–9.

Tuckwell, H. C. (1985). Some aspects of cable theory with synaptic reversal potentials. *J. Theoret. Neurobiol.* **4** 113–27.

Tuckwell, H. C. (1986a). On shunting inhibition. *Biol. Cybernetics* **55** 83–90.

Tuckwell, H. C. (1986b). Stochastic equations for nerve membrane potential. *J. Theoret. Neurobiol.* **5** 87–99.

Tuckwell, H. C. (1987a). Diffusion approximations for channel noise. *J. Theor. Biol.* **127** 427–38.

Tuckwell, H. C. (1987b). Statistical properties of perturbative nonlinear random diffusion from stochastic integral representations. *Phys. Lett. A* **122** 117–20.

Tuckwell, H. C. (1987c). Perturbative analysis of random nonlinear reaction–diffusion systems. *Physica Scripta* **36** (In press).

Tuckwell, H. C. (1988a). *Elementary Applications of Probability Theory*. Chapman and Hall, London.

Tuckwell, H. C. (1988b). *Stochastic Processes in the Neurosciences*. SIAM, Philadelphia.

Tuckwell, H. C., and Cope, D. K. (1980). Accuracy of neuronal interspike times calculated from a diffusion approximation. *J. Theor. Biol.* **83** 377–87.

Tuckwell, H. C., and Hermansen, C. E. (1981). Ion and transmitter movements in spreading cortical depression. *Int. J. Neurosci.* **12** 109–135.

Tuckwell, H. C., and Miura, R. M. (1978). A mathematical model for spreading cortical depression. *Biophys. J.* **23** 257–76.

Tuckwell, H. C., and Richter, W. (1978). Neuronal interspike time distributions and the estimation of neurophysiological and neuroanatomical parameters. *J. Theor. Biol.* **71** 167–83.

Tuckwell, H. C., and Walsh, J. B. (1983). Random currents through nerve membranes. I. Uniform Poisson or white noise current in one-dimensional cables. *Biol. Cybernetics* **49** 99–110.

Tuckwell, H. C., and Wan, F. Y. M. (1980). The response of a nerve cylinder to spatially distributed white noise inputs. *J. Theor. Biol.* **87** 275–95.

Tuckwell, H. C., and Wan, F. Y. M. (1984). First passage time of Markov processes to moving barriers. *J. Appl. Probab.* **21** 695–709.

Tuckwell, H. C., Wan, F. Y. M., and Wong, Y. S. (1984). The interspike interval of a cable model neuron with white noise input. *Biol. Cybernetics* **49** 155–67.

Uhlenbeck, G. E., and Ornstein, L. S. (1930). On the theory of Brownian motion. *Phys. Rev.* **36** 823–41.

Van der Kloot, W., Kita, H., and Cohen, I. (1975). The timing and appearance of miniature end-plate potentials. *Prog. Neurobiol.* **4** 269–326.

Van Dobben de Bruyn, C. S. (1968). *Cumulative Sum T[illegible]*. Griffin, London.

Vasudevan, R., and Vital, P. R. (1982). Interspike interval density of the leaky integrator model neuron with a Pareto distribution of PSP amplitudes and with simulation of refractory time. *J. Theoret. Neurobiol.* 1 219–27.

Velikaya, R. R. and Kulikov M. A. (1966). Statistical analysis of impulse activity of neurones of the cerebral cortex. *Biofizika* 11 321–8.

Verveen, A. A., and DeFelice, L. J. (1974). Membrane noise. *Prog. Biophys. Mol. Biol.* 28 189–265.

Volle, R. L., and Brainsteanu, D. D. (1976). Quantal parameters of transmission at the frog neuromuscular junction. *Arch. Pharmacol.* 295 103–8.

Walloe, L., Jansen, J. K. S., and Nygaard, K. (1969). A computer simulated model of a second order sensory neurone. *Kybernetik* 4 130–45.

Walpole, R. E., and Myers, R. H. (1985). *Probability and Statistics for Engineers and Scientists*. MacMillan, New York.

Walsh, J. B. (1981a). Well-timed diffusion approximations. *Adv. Appl. Probab.* 13 352–68.

Walsh, J. B. (1981b). A stochastic model of neuronal response. *Adv. Appl. Porobab.* 13 231–81.

Wan, F. Y. M., and Tuckwell, H. C. (1979). The response of a spatially distributed neuron to white noise current injection. *Biol. Cybernetics* 33 39–55.

Wan, F. Y. M., and Tuckwell, H. C. (1982). Neuronal firing and input variability. *J. Theoret. Neurobiol.* 1 197–218.

Washio, H. M., and Inouye, S. T. (1980). The statistical analysis of spontaneous transmitter release at individual junctions on cockroach muscle. *J. Exp. Biol.* 87 195–201.

Watanabe, S., and Creutzfeldt, O. D. (1966). Spontane postsynaptische potentiale von nervenzellen des motorischen cortex der katze. *Exp. Brain. Res.* 1 48–64.

Waxman, S. G. (1977). Conduction in myelinated, unmyelinated and demyelinated fibers. *Arch. Neurol.* 34 585–9.

Webb, A. C. (1976a). The effects of changing levels of arousal on the spontaneous activity of cortical neurones. I. Sleep and wakefulness. *Proc. Roy. Soc. Lond. B* 194 225–37.

Webb, A. C. (1976b). The effects of changing levels of arousal on the spontaneous activity of cortical neurones. II. Relaxation and alarm. *Proc. Roy. Soc. Lond. B* 194 239–51.

Westerfield, M., Joyner, R. W., and Moore, J. W. (1978). Temperature sensitive conduction failure at axon branch points. *J. Neurophysiol.* 41 1–8.

White, B. S., and Ellias, S. (1979). A stochastic model for neuronal spike generation. *SIAM J. Appl. Math.* 37 206–33.

Whitsel, B. L., Roppolo, J. R., and Werner, G. (1972). Cortical information processing of stimulus motion on primate skin. *J. Neurophysiol.* 35 691–717.

Whitsel, B. L., Schreiner, R. C., and Essick, G. K. (1977). An analysis of variability in somatosensory cortical neuron discharge. *J. Neurophysiol.* 40 589–607.

Wilbur, W. J., and Rinzel, J. (1982). An analysis of Stein's model for stochastic neuronal excitation. *Biol. Cybernetics* 45 107–15.

Wilbur, W. J., and Rinzel, J. (1983). A theoretical basis for large coefficient of variation and bimodality in neuronal interspike interval distributions. *J. Theor. Biol.* 105 345–68.

Wilson, D. M., and Wyman, R. J. (1965). Motor output patterns during random and rhythmic stimulation of locust thoracic ganglia. *Biophys. J.* 5 121–43.

Woodward, D. J., Hoffer, B. J., and Altman, J. (1974). Physiological and pharmacological properties of Purkinje cells in rat cerebellum degranulated by postnatal X-irradiation. *J. Neurobiol.* 5 283–304.

Wyler, A. R., and Fetz, E. E. (1974). Behavioral control of firing patterns of normal and abnormal neurons in chronic epileptic cortex. *Exp. Neurol.* 42 448–64.

Wylie, R. M., and Pelpel, L. P. (1971). The influence of the cerebellum and peripheral somatic nerves on the activity of Deiters' cells in the cat. *Exp. Brain Res.* **12** 528–46.

Wyman, R. J. (1966). Multistable firing patterns among several neurons. *J. Neurophysiol.* **29** 807–33.

Yana, K., Takeuchi, N., Takikawa, Y., and Shimomura, M. (1984). A method for testing an extended Poisson hypothesis of spontaneous quantal transmitter release at neuromuscular junctions. *Biophys. J.* **46** 323–30.

Yanagida, E. (1985). Stability of fast travelling pulse solutions of the Fitzhugh–Nagumo equations. *J. Math. Biol.* **22** 81–104.

Yang, G. L., and Chen, T. C. (1978). On statistical methods in neuronal spike train analysis. *Math. Biosci.* **38** 1–34.

INDEX